Prof. Thorpe

Coal

Its History and Uses

Prof. Thorpe

Coal
Its History and Uses

ISBN/EAN: 9783337059620

Printed in Europe, USA, Canada, Australia, Japan

Cover: Foto ©berggeist007 / pixelio.de

More available books at **www.hansebooks.com**

COAL

ITS HISTORY AND USES

BY PROFESSORS

GREEN, MIALL, THORPE, RÜCKER, AND MARSHALL

Of the Yorkshire College

EDITED BY PROF. THORPE

London

MACMILLAN AND CO.

1878

PREFACE.

THE ORIGIN of this book may be told in a very few
words. Towards the close of last year Dr. Carpenter,
as Secretary to the Gilchrist Educational Trust, suggested to the Professors of the Yorkshire College that
by delivering courses of lectures in connection with
the Trust, in some of the larger towns of the West
Riding, they might aid in the good work which is
being done by means of Dr. Gilchrist's bequest. No
special conditions were attached to the proposal. Each
lecturer was to be free to choose his own subject, consistent, of course, with the general provisions of the
Trust. It was thought, however, that the educational
value of the lectures would be increased by selecting
some common and familiar subject, sufficiently comprehensive to allow each lecturer to treat it in a
manner appropriate to his own studies and to the
teaching department which he represents in the
College. Various reasons conduced to the selection of
the particular subject of Coal. It was not too ambi-
tious ; it was sufficiently familiar and sufficiently comprehensive ; and, above all, it had a direct, practical,

every-day interest for the class to which the lectures
were addressed. Ten lectures on Coal were accordingly delivered in Leeds and Keighley early in the
present year. Local committees undertook to provide
suitable rooms and to announce the course in each
town ; and their prompt and thorough co-operation
largely promoted the success of the enterprise.

In the preparation of the lectures much information
of a kind not hitherto brought before the general public
was gathered together, often from ·sources not readily
available to ordinary readers. The publication of the
substance of the course was suggested, in the hope that
a volume on Coal might interest a wider public than
that originally addressed. This book, then, consists of
the lectures, altered in form, revised, and arranged
consecutively in the form of chapters. Chapters I.
and II., on the Geology of Coal, are by Professor
Green ; Chapter III., on Coal-plants, and Chapter IV.,
on the Animals of the Coal-measures, are by Professor
Miall ; Chapters V. and VI., on the Chemistry of Coal,
are by Professor Thorpe ; Chapters VII. and VIII., on
Coal as a Source of Warmth and Power, are by Professor Rücker ; and Chapters IX. and X., on the Coal
Question, are by Professor Marshall.

A generation has passed away since Joule first
determined the Mechanical Equivalent of Heat ; but
the doctrine of the Correlation of Heat and Work has

hardly yet come down to the popular intelligence :
nay, three generations have elapsed since Erasmus
Darwin maintained that Coal was formed out of ancient
morasses and forests, but what proportion of those
who use Coal in these latter days know anything of
the mystery of its origin ? If the knowledge of these
things is in future to spread at no greater rate than this,
it is to be feared that our Coal will be at an end before
our people have learned to know what it actually is,
how it has been formed, and what it can do.

CONTENTS.

ILLUSTRATIONS.

C O A L.

CHAPTER I.

THE GEOLOGY OF COAL.

COAL, looked at from three different standpoints, is to form the subject of this book.

We propose to go back into the past, that we may sketch out the state of the country at the time when coal was coming into being, and the processes by which it was formed, to describe, too, the plants and animals then living; next, to deal with the present and give an account of some of the uses to which coal is now being put; lastly, if it may be, to forecast the future and speak of the probable duration of our coal supply.

It falls naturally to the lot of the geologist to begin the story. His part is that of an historian, who promises to tell what the part of the world we live in was like, and what was going on in it, during the time when our coal was being formed; a time so remote that, compared with its incidents, the most distant events that history depicts or tradition shadows forth are as things but of yesterday.

Has the thought already crossed the minds of any of my readers, that I must be a bold man to make such a promise? Are any of them asking where I am to get my information from, and what grounds I can have for pretending to know anything whatever about an epoch so far removed from our day ? Has a suspicion arisen that my narrative, though possibly plausible enough, will be mainly a product of the imagination, with as little solid foundation of fact to rest upon as a fairy tale ?

I am free to confess that in the execution of a task such as I have undertaken imagination must have a place, but at the same time, I confidently assert, no larger share than in the compilation of the most orthodox and trust-worthy histories.

I need scarcely say that to string together a number of extracts from chronicles giving an account of what is said to have happened in bygone times, is not to write history. Before its statements can be accepted, the accu-racy of the chronicle must be inquired into; and here nothing aids the historian so powerfully as an active imagination, for by its help he can put himself in the place of the chronicler, can see where his judgment was liable to be warped and his eyesight dimmed by prejudice and misconception, and where there is even a risk of his having been guilty of deliberate misrepresentation.

Just so much and no more has imagination influenced the compilation of the history I am about to unfold ; I have striven to carry myself back to the conditions and circumstances of the period with which we have to deal, and knowing what would be the result of such conditions and circumstances at the present day, I infer that similar results followed from them in bygone times.

All well and good, perhaps some one will say, but will the geologist have the kindness to inform us how he knows what were the conditions and circumstances of that far distant period ?

It will be my main object in the present chapter to answer this question, and to explain how geologists acquire a knowledge of the physical conditions that prevailed during past periods of the earth's lifetime.

The records from which geologists draw their information can be scarcely compared to written or printed histories. There are however nations of whom no written account exists, who perhaps never had any written history, but about whom we are still able to gather from other sources a vast amount of information. Their houses, their monuments, their weapons, and their tools have survived, and these tell us of the kind of life, the state of civilisation, and the skill of the men to whom they belonged; from the contents of their tombs we learn what manner of men they were physically; sometimes a sudden change in the appointments and belongings of the folk indicates that tribes which had for long inhabited a district were driven out and replaced by a new race. Thus from waifs and strays we can piece together a fairly connected account of the events of a period long antecedent to any written history.

The investigations of Dr. Schliemann on the supposed site of the city of Troy furnish a good example of this method of research. He found lying one on the top of another, traces of the existence of five successive communities of men, differing in customs and social development, and was able to establish the fact that some of the cities had been destroyed by fire, and that later on other towns had grown up over the buried remains of the earlier settlements. The lowest layers were of course the oldest, and the position of each successive layer in the pile gives its date, not in years but with regard to the layers above and below it.

Now from time immemorial nature has been at work building up monuments and providing tombs, which tell us what were the events going on and what kind of inhabitants the earth had long before man made his appearance on its surface. The monuments are the rocks which compose the ground under our feet, and these, like many ancient monuments of human construction, are the tombs of the creatures that lived while they were being built.

Many facts testify that the earth's crust did not come into existence exactly as we find it now, but that its rocks have been built up by the slow action of natural agencies. These rocks constantly enclose the remains of plants and animals, and as it is evident that neither plant nor animal could have lived in the heart of a solid rock, this fact shows that the rock must in some way have gathered around the remains which are now found in it. Again, many of these fossils belonged to animals that lived in water, the larger part indeed to marine creatures. This indicates that the rock was formed beneath the sea, and when we examine the way in which the constituents of the rock are arranged, we frequently find it to correspond exactly with the manner in which the sand and mud that rivers sweep down into the sea or lakes are spread out over the bottom of the water. In a pile of rocks formed in this way it is clear that the lowest is the oldest of all, and that any one stratum is younger than that which lies beneath it and older than that on the top of it. Further the occurrence of rocks inland containing marine fossils far above the sea level shows that sea and land have changed places. When again we find that the fossils of one group of rocks differ totally from those of a group lying above, we learn that one race of creatures died out and was supplanted by a new assemblage of animal forms.

These general remarks will, I trust, give some notion of the evidence which is available for reconstructing the history of those remote periods with which geology deals, and of the kind of reasoning which the geologist employs for interpreting the records that are submitted to him. We will now examine in detail, by the aid of these methods, the group of rocks among which coal occurs in Britain, and see how far we can read the story they have to tell.

The group with which we have to deal is called the Carboniferous or Coal-bearing System, and it includes four classes of rocks, viz. :

1. Sandstone.
2. Shale or Bind.
3. Limestone.
4. Coal and Under-clay.

We will take the Sandstones and the Shales first. They scarcely require description, but to avoid any risk of misconception it will be as well to say that sandstone consists of grains of sand, and that sand is nothing else but particles of a very hard mineral known to mineralogists as quartz, and consisting of a substance called silica by chemists.[1] The grains are bound together by a cement which is in some few cases identical in composition with themselves and consists of pure silica, but usually is a mixture of sandy, clayey, and other substances. Sometimes the grains are small and all pretty much of a size, and then we have a finely grained sandstone; sometimes some of the grains are considerably larger than others, the broken surface is then rough to the touch, and the rock is called a gritstone; sometimes pebbles of quartz are embedded in a mass of sandy grains, and form a conglomerate or puddingstone.

The shales are made up very largely of clay,[2] mixed however usually with sand or other substances.

---

[1] The chemical composition of silica is SiO_2.

[2] Pure clay is a hydrated silicate of alumina with the composition, $2SiO_2. Al_2O_3 + 2H_2O$. It is formed by the decomposition of potash felspar, whose composition is, $6SiO_2. Al_2O_3. K_2O$. Clayey rocks consist of pure clay mixed with sand and other foreign substances.

Both sandstones and shales are divided into layers or beds, and are said to be stratified.

It is this stratified or bedded structure that gives us the first clue to the way in which these rocks were formed. Rivers are constantly carrying down sand and mud into the sea or lakes, and when their flow is slackened on entering the still water, the materials they bring down with them sink, and are spread out in layers over the bottom. If at any time there is a pause in the supply of sediment, the layer last deposited hardens before the next is laid upon the top of it. Thus a plane of separation between the two layers is formed, and by a repetition of this process the deposit comes to be divided into layers or strata. The stratified structure of the sandstones and shales shows that they were formed in this way; they often enclose remains of plants that have been carried down from land, and occasionally of animals that lived in the water where they were deposited.

But though sandstones and shales agree in being rocks formed under water, there are important differences between them. The material, quartz, of which sandstones are composed is very hard, and will stand a great deal of rolling and knocking about without being reduced to a fine state of division. The sandy sediment then which rivers carry down will tend to be coarse, and therefore heavy; it will require a powerful current to sweep it along, and when the current that moves it has its velocity checked by entering a body of still water, it tends to fall quickly to the bottom and to be piled up more or less in heaps. Clay, on the other hand, is a soft substance, easily reduced to a fine state of division, which allows it to float suspended in water, and it can be carried forward by currents so gentle as to be quite incapable of moving the heavy grains of sand. A river, therefore, coming down with a strong current, though it must experience some check on entering the sea, will retain for long distances velocity enough to enable it to carry forward finely divided mud, and sediment of this character will settle down very slowly, and will be spread over the bottom in uniform and very extensive layers.

The forward motion of the current will scarcely affect a grain of sand, and it will fall almost vertically to the bottom. The path described by a particle of fine clay will result from two combined movements: gravity will slowly carry it down, and at the same time it will be sensibly swept on in a horizontal direction by the current; it will move in fact as if it were rolling down an inclined plane with a gentle slope. The resistance of the water will somewhat modify these results. In bodies of similar shape the resistance increases with the extent of surface exposed, and will therefore be greater in the case of rounded grains of sand than for particles of fine clay; but the surface, and therefore the resistance, increases as the square of the diameter, and the bulk or weight as the cube, so that even when the retarding effect of the resistance is taken into account, sand will still descend faster than clay. Mr. Sorby finds that grains of sand $\frac{1}{100}$ of an inch in diameter subside a foot in ten seconds, while the smallest grains of kaolin (pure clay) take about five days to descend through the same space. He adds however that in nature a complete separation of clay from sand seldom occurs. The particles of kaolin have a very great tendency to stick together and collect into complex granules, and in so doing may enclose grains of sand.[1] It is doubtless owing to this tendency that so many of the shales contain a large admixture of sand.

Suppose, then, we have a river which is bringing down

[1] *Monthly Microscopical Journal*, 1877, Anniversary Address, p. 15.

into the sea sandy and clayey sediment of all degrees of grain, from large pebbles to the most impalpable mud, it is easy to see what will be the law which governs the distribution of the deposits laid down on the sea bed.

The least check will suffice to bring the heaviest and coarsest materials to rest. The pebbles and the coarsest sand will be thrown down close to the shore in banks, having the form of a wedge with a steep seaward face and a large vertical angle, such as A in fig. 1.

Sand a little less coarse will travel a little further, and form a bank with a face not quite so steep and with a smaller vertical angle, such as B in fig. 1. And so the arrangement of the sediment will go on; as we recede from the shore, the material will grow finer and finer, and will contain less and less sand and more and more clay, the wedge-shaped form of the banks will become less and less pronounced, till at last the vertical angle will become so small that the upper and under surfaces will become sensibly parallel, and the bank will pass into a bed. These gradations are shown in a diagrammatic form in fig. 1, where the dotted portions represent sandy deposits, the coarseness of the grain being indicated by the size of the dots, and the clayey deposits are distinguished by fine parallel lines.

The lesson that we learn from these facts is this. Supposing that we find along a certain line of country coarse, sandy, and conglomeratic rocks to prevail, and that as we travel away from this line the beds grow finer in grain, less and less sandy, and gradually pass into regularly stratified clays; we may infer with absolute safety, that, at the time these rocks were being formed, the belt where the coarse sandy rocks are in force, ran parallel to the coast line and at no great distance from it,

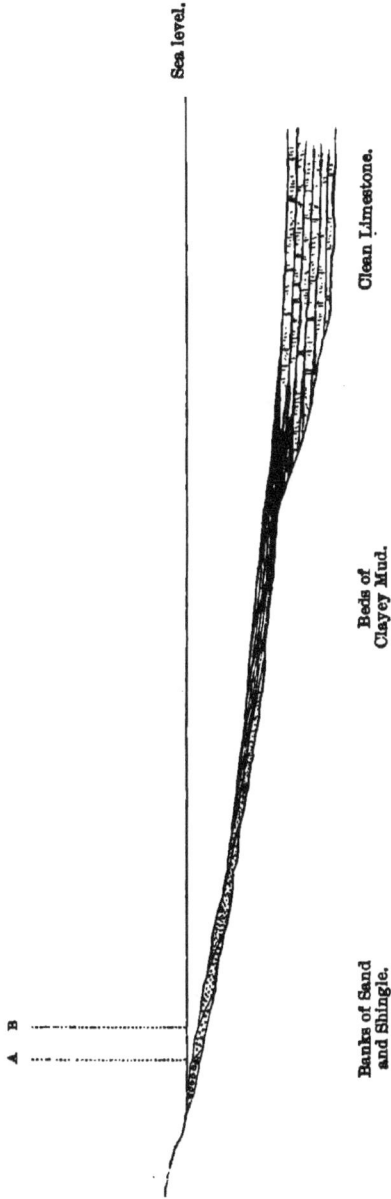

FIG. 1.

Sea level.

Clean Limestone.

Beds of
Clayey Mud.

Banks of Sand
and Shingle.

A   B

and that the sea spread out towards the quarter in which the rocks grow gradually finer and more clayey.

Here, then, a study of the composition and character of a group of rocks enables us to map out the distribution of land and sea at the time they were being formed, and this instance will give some notion of the method which geologists employ to arrive at a solution of this and similar problems.

Before leaving this part of the subject, I would once more lay stress upon the marked difference between the ways in which coarse sandy and fine clayey material will accumulate on the sea bottom. The coarse, heavy stuff falls down in heaps, and forms wedge-shaped masses. This is exactly the character which beds of sandstone are found to possess; they never keep for far the same thickness, and frequently can be actually observed growing thinner and thinner till they taper out. In fact, it is scarcely correct to speak of *beds* of sandstone at all; sandy rocks occur nearly always in more or less wedge-shaped banks. The fine clayey material, on the other hand, settling down slowly is spread with great regularity over large spaces in beds of constant thickness. This is exactly the case with shale. It is finely and evenly bedded, splitting into thin regular plates, and it occurs in masses which keep the same thickness over large areas. Here then the results which theory predicts are just those which observation shows have been brought about.

The next rock we have to consider is Limestone. Limestone is mainly made up of a substance known to chemists as Calcium Carbonate, or Carbonate of Lime. Most limestones however contain besides clay, sand, or other impurities in larger or smaller quantity.

There is one point I must make clear, if I am to explain the way in which limestone was formed. Many limestones, the marble for instance used for statuary, are extremely hard and solid rocks; but solid and hard as they are, these limestones can be dissolved by water as surely, though not so fast, as salt or sugar. Pure water cannot produce this effect, but water holding carbonic acid[1] in solution, has the power of dissolving the carbonate of lime that enters so largely into the composition of limestone. Nearly all the water on the earth's surface holds carbonic acid dissolved in it, the gas being drawn to a slight extent out of the air by rain, and obtained much more largely from decaying organic matter; and therefore nearly all water has, and exercises whenever opportunity occurs, the power of dissolving carbonate of lime out of limestone.

Proofs of the dissolution of limestone are not hard to meet with. Go to a spot where a stream which has been running for some distance over limestone falls in a small cascade or trickles down a rough steep bank. Everything on which the water breaks, stones, twigs, blades of grass, is coated with a white friable deposit, which on examination is found to be carbonate of lime. The substance was taken up by the stream in its passage over the limestone and carried down in solution; when the water is scattered in spray at the little fall or rapid, a portion of the carbonic acid is dissipated by diffusion, and the dissolved carbonate of lime is deposited in a solid form. Not to go far from home, Derbyshire and the country round Clapham will furnish numerous instances; there is a very good example near Miller's Dale Station, and another in the upper part of the waterfall at Gordale Scar.

[1] The chemical composition of carbon dioxide is CO_2, carbonic acid is a compound of carbon dioxide and water, with the composition, H_2CO_3.

In some districts, specially in volcanic countries, springs occur very highly charged with carbonate of lime, and by their agency the bulk of carbonate of lime carried down in a dissolved state by rivers is largely increased. The warm, so-called petrifying, springs of Matlock are a case in point; they are probably the last vestiges of volcanic action which was in operation in that neighbourhood during Carboniferous times.

Hence, whenever water runs over limestone, it takes up and carries away in an invisible form into the sea more or less carbonate of lime, and when calcareous springs occur, they largely swell the amount of dissolved matter. In this way we obtain the material for the formation of submarine limestones; the next step is to explain how the carbonate of lime is got out of the water so as to be available for the formation of rock. This is effected by the agency of certain animals which have the power of extracting the dissolved carbonate of lime, and using it to build up the hard parts of their bodies or the dwellings in which they live. The most important of the animals that do this work are minute creatures of low organisation called Foraminifera, shell fish, and corals. In the case of some of the carboniferous limestones, sea-lilies, or crinoids, have played an important part in this work. By the aid of these animals, the carbonate of lime is brought back to a solid form; at their death their hard parts fall to the bottom, and accumulate in a mass which has the composition of a very pure limestone, and which afterwards becomes solidified into limestone-rock.

If the accumulation of carbonate of lime goes on in a part of the sea where mud or sand is falling through the water, the result is an impure limestone containing admixtures of clay or sand. In such regions too we shall find alternations of limestone, usually impure, with sandstones and shales. As long as the state of the water was such as to allow of limestone-forming creatures living in it, the deposit on the bottom consisted of a mixture of their hard parts with sand and mud; every now and then however the water became so fouled with sediment that the limestone-formers were either killed or driven away, and then only sand or mud was deposited. When the water became clearer, the animals returned and the growth of limestone was resumed.

There are however portions of the ocean bed on to which neither mud nor sand finds its way. Even the most finely divided clay, slowly as it settles down, must at last reach the bottom, and there will therefore be a boundary which no suspended material crosses. On the seaward side of this line, the seafloor receives no mechanically borne sediment; but the carbonate of lime which is carried in solution can travel to any distance, and provided the requisite animals be present, the formation of limestone may go on in any part of the ocean. And it is in those regions which are too far from land to allow of mechanically carried sediment reaching them, that the building up of limestones by organic agency goes on most vigorously. The limestone of these tracts will differ from that formed nearer the shore in being composed of very little else but carbonate of lime, and in having no shales or sandstones interstratified with it.

That this explanation of the formation of limestones is sound and good, we have a twofold proof.

Many of the limestones of the earth's crust can be seen even now to be made up almost entirely of the hard parts of marine animals. The microscope for instance tells us that chalk is often little else but a slightly compacted mass of the tiny shells of foraminifera; and parts of the Carboniferous Limestone, which are often fashioned into mantel-pieces, are a complete mash of the broken stems of sea-lilies. There are limestones, it is true, in which no trace of organic remains can now be detected; but we can often trace a passage from such rocks through others, in which fossils are first faintly indicated, and then become more and more distinct, into limestones whose organic origin is palpable to the most casual observer. We are therefore justified in concluding that the whole suite had a common origin, but that in certain parts the fossils have been effaced by changes which the rock has undergone since its formation.

Again we know that the process which gave rise to limestone long ago, is still at work. The deep-sea explorations of recent years have shown that a deposit, which differs in no respect from earthy chalk, is forming over vast portions of the profound abysses of the ocean by the gradual accumulation of shells of foraminifera; and in the Pacific the minute coral-polyps are slowly building up vast piles, which, if they are ever raised above water, will rank among the most important limestones of the future.

The information then that limestones give us about ancient seas is this:—When we find, as we often do, a mass of limestone, hundreds of feet thick and composed of little else but carbonate of lime, we know that the spot where it occurs was, at the time it was formed, far out at sea, covered by the clear water of mid-ocean; and when we find that this limestone grows in certain directions earthy and impure, and that layers of shale and sandstone, thin at first, but gradually thickening out in a wedge-shaped form, come in between its beds, we know that in those directions we are travelling towards the shore lines of that sea, whence the water was receiving from time to time supplies of muddy and sandy sediment.

And now perhaps it will be admitted that I was not so very bold after all, when I promised to tell what this part of the world was like during the far distant epoch when coal was formed, for I shall certainly have made a step in that direction if I can show how to determine the way in which land and sea were then distributed.

If I have been successful in my explanation, it will now come home to us that we may have in a sheet of rock a picture of the character and extent of the old sea in which that rock was formed, as graphic and as easy to understand as that which a hydrographical chart furnishes of the character and extent of a modern ocean. The hydrographer denotes by certain conventional colours and signs the depth and nature of the bottom of different parts of the ocean. On the great charts which nature has provided for us of the oceans of bygone days, the very same information is given by the mineral character and fossil contents of different parts of the same sheet of rock.

We find that the middle of this sheet of rock consists of clean limestone; that around this central patch the limestone grows gradually more and more earthy and shades away into shale; the shale in its turn, as we travel outwards from the centre, becomes more and more sandy and merges into sandstone; the sandstone grows coarser and coarser in grain till we reach an outer belt of conglomerate.

The chart has an index which explains to us the meaning of the colours and signs employed on it, and the train of reasoning which we have been following enables us to draw up an index on which each of the different forms assumed by the sheet of rock is made to indicate the part of the sea bed on which it was deposited.

The belt of pebbly conglomerate defines the margin of that sea, as clearly as the conventional shading used on -

charts to denote the zone of shallow water that fringes
the coast. The gradual passage into finer and more
clayey forms of rock, tells of increasing distance from the
shore; and the central patch of clean limestone marks
out a tract so far from land that no sediment can reach it
to foul the deep blue of its limpid water.

Fig. 1, in which these truths are shown in a diagrammatic form, is in fact such an index as I have spoken of.

And now we come to coal, a rock which constitutes but
a small portion of the whole bulk of the carboniferous
deposits, but which may yet be fairly looked upon as the
most important member of that group, both on account of
its intrinsic value and also from the interest that attaches
to its history.

That coal is little else but mineralised vegetable
matter is a point on which there has been for a long time
but small doubt. The more minute investigations of
recent years have not only placed this completely beyond
question, but have also enabled us to say what the plants
were which contributed to the formation of coal and in
some cases even to decide what portions of those plants
enter into its composition.

The chemical composition of coal renders its vegetable
origin, to say the least, probable. It is composed of carbon
oxygen, and hydrogen, with a small proportion of nitrogen;
it consists therefore of the elements which make up all
vegetable organic compounds.

It is true that the chemical composition of coal differs
more or less, in the case of some coals differs very widely,
from that of woody fibre; but considerations to be hereafter explained show that this does not in any way forbid
the belief that coal is of vegetable origin.

Chemical composition alone then would furnish a
strong presumption that coal had a vegetable origin; and

this inference is fully confirmed when we find that a large portion of most coals can be seen even now to be made up of vegetable structures.

The measures in which coal occurs are mainly made up of sandstones and shales, and the coal lies among these rocks in seams or layers, which run parallel to their bedding. If we go into a coal pit, or an open cutting where a seam of coal is exposed, and break down some of the coal, we find that it splits most easily in three directions nearly at right angles to one another, so that it comes away in rudely cubical masses. Two of these directions are roughly at right angles to the planes of bedding of the rocks among which the coal occurs. The faces of the block of coal on these sides are smooth and shining and do not soil the fingers; one of the faces called the 'bord' or 'cleat' is very marked, the other called the 'end' is less sharply defined. The third direction in which the coal naturally divides is parallel to the bedding of the rocks above and beneath it. The planes of division in this direction are dull and grimy to the touch, owing to a thin layer or numerous patches of a soft black substance which looks exactly like charcoal and is called 'mineral charcoal' or 'mother of coal.' The woody character of this mother of coal is palpable even to the unaided eye, and when it is examined under the microscope the minute tissues and structures are seen to be so perfectly preserved that it is possible in some cases to say what the plant was of which they originally formed a part.

In the solid lustrous portion of the coal traces of vegetable tissue are not obvious to the unaided eye, but they may be detected even here by careful scrutiny or microscopic examination.

On viewing the surfaces of the laminæ of some Nova Scotian coals under a strong oblique light, Dr. Dawson

observed them to be impressed with the markings which
characterise several plants found fossil in plenty in the
Coal Measures, and in some cases they presented the outline of flattened trunks. In the roof-shales similar
flattened coaly trunks are more easily recognised, because,
instead of composing the whole mass of the deposit as in
coal, they are separated from one another by clay. These
trunks consist of an external layer of coal enclosing in
some cases a film of mineral charcoal. That the coal has
been formed out of the bark, and that the mineral charcoal
is all that is left of the wood appears likely for the following reasons. Stems of trees, especially of one called
Sigillaria, are occasionally met with in the Coal Measures
standing erect as they grew. These stems are now columns
of shale or sandstone surrounded by a thin, cylindrical shell
of bright, clean coal; the central portion often contains
leaves, shells, and other vegetable or animal remains, and
sometimes a small pile of mineral charcoal has been
noticed at the bottom. The tree therefore must have
become hollow before it was embedded; the internal
woody cylinder rotted away, leaving a shell of bark. Sand
and mud were deposited around; at the same time the
wood disappeared under the influence of decay or broke up
and bit by bit rose to the surface of the water and floated
away, and its place was taken by sediment, with which
were mixed fragment of plants and the remains of animals
that had taken shelter in the hollow trunk; the bark
then became converted into coal and the trifling remnant
of the wood into mineral charcoal. In a few cases
the stem preserved its upright position; in others the
bark fell in and all that remains to indicate where a tree
once stood is a little heap of mineral charcoal with strips
of coaly bark; usually the trunk toppled over and was
squeezed flat by the weight of the sediment afterwards

deposited above it. The supposition that the purer
portions of the coals which Dr. Dawson studied are mainly
bark is confirmed by the fact that, when they were examined under the microscope, the traces of structure
which were sufficiently well preserved to be capable of
recognition agreed with the structure of the bark of
Sigillaria.

It should be mentioned too that the total removal of
the wood of a dead tree leaving the bark intact is a process which goes on now-a-days, especially in hot countries.
In tropical forests the traveller constantly comes across
what appear at first sight to be solid fallen trunks, but on
stepping upon them, the foot goes through a thin shell
into a hollow space swarming with inhabitants of not the
most agreeable character : the wood has all gone, and a
thin cylinder of bark alone remains. Dr. Dawson does
not assert that all the coals examined by him consist of
fossil charcoal and mineralised bark. In the coarser
seams remains of herbaceous plants and leaves are present
in large quantities, and some small beds appear to be
wholly made up of such materials.

These Nova Scotian coals then seemed to consist of
three portions. Mineral charcoal, which is wood from
which the oxygen and hydrogen have been to a great
extent abstracted leaving a carbonised residue in which
the original vegetable structure is very apparent; clean
lustrous coal, mainly made up of bark and in which the
structure has been very largely obliterated; and the
coarser coal, composed in large measure of leaves and
other portions of herbaceous plants.

The examination of our English coals has occupied the
attention of many microscopists, and their investigations
have placed it beyond question that many of these coals
are composed of parts of plants very different to those

which make up the bulk of the Nova Scotian coals. It is now very generally admitted that many English and other coals are little else but an aggregation of minute rounded bodies, known as 'spores,' that were shed by a tree called Lepidodendron, which is frequently found fossil in the Coal Measures.

This plant and its modern representatives will be fully described in a subsequent chapter, but a short account of it must be given here.

The Lepidodendron, or scaly-tree, is so called because it was covered outside by lozenge-shaped markings, which bear some resemblance to the scales of a fish. It reached a considerable size, spread out atop into forked branches, and to the ends of these there were attached conical fruits called Lepidostrobus. It is the character of these cones which enables us to fix the place of Lepidodendron in the vegetable kingdom. Their structure clearly shows that it belongs to the order Lycopodiaceæ, the best known member of which now-a-days is the club-moss.

The evidence by which this conclusion is reached is as follows.

The reproductive organs of Lycopodiaceæ are contained in spikes or cones that consist of overlapping scales or leaves. In the space between each two leaves, there is a bag called a 'sporangium,' filled with spores. In some genera the spores are of two sizes, and the larger are distinguished as 'macrospores' and the smaller as 'microspores.' In other genera all the spores are of the same size.

It is unnecessary to explain by what steps these spores effect the multiplication of the plants, but it is important to notice that they are of a highly resinous and inflammable nature. A pinch of Lycopodium spores thrown into the air and touched with a taper is instantly consumed with a sudden flash, which mimics lightning closely enough to have been used for that purpose on the stage. The resinous character also prevents the spores from being wetted when they come in contact with water, and tends to preserve them from decay.

Now the fruit of Lepidodendron resembles in every essential the spikes or fertile branches of some modern Lycopods. Lepidostrobus in well preserved specimens is seen to consist of overlapping scales which support sporangia. The sporangia contain spores, and on the under side of these there is a three-rayed marking, which is caused by four spherical spores having been pressed close together. All the spores of Lepidostrobus are of the same size, but in another fossil cone called *Triplosporites*, which doubtless belonged to some lepidodendroid tree, both macrospores and microspores are present.

But though Lepidodendron was unquestionably a lycopod, it presents in one respect a striking contrast to all living members of that order. Our English lycopods are diminutive plants, and the largest foreign species rise only to the height of a few feet, but Lepidodendron was a forest tree, and may have attained in some cases a height of a hundred feet.

And now for the bearing of this on the composition of English coals. As an example, take the Better Bed coal of Bradford. A block of this consists of layers of bright clear coal parted by films of mineral charcoal. The mineral charcoal agrees in character and composition with that of the Nova Scotian coals, the difference between these and the Better Bed lies in the solid black layers. When a slice of this part, cut parallel to the bedding, is ground thin and examined under the microscope, it is found to consist of two parts. There is a black or dark brown granular ground-mass, and scattered through this ground-mass there are numerous semi-transparent discs and rings of a yellow colour. In a section cut perpendicular to the bedding, elongated yellow bars run through the ground-mass roughly parallel to the bedding, and some of them are like vertically flattened hoops. It is clear that these yellow patches are round bags that have been flattened by pressure. On some of the discs a three-rayed marking, corresponding exactly to that on the side of the spores of Lepidostrobus can be recognised, and the spores in the cone and patches in the coal also agree in size. There can be no doubt then that these yellow particles are Lepido-dendron spores.

The sacs just described average about one-twentieth of an inch in diameter. Inside them, and occasionally scattered through the ground-mass, very much smaller translucent yellow spots have, it is said, been detected in some cases. The probability is that the larger bags are macrospores and the smaller spots microspores.

We know then where the mineral charcoal and the larger yellow patches came from. What is the dark ground-mass? Professor Huxley believes it to be highly carbonised spores. He says that in coals burning with flame the visible spores are very numerous, and that from such coals we can trace a passage through others, in which the spores become more and more indistinct and the ground-mass gradually increases in amount, into An-thracite, in which the ground-mass predominates to such an extent that it is impossible to grind a section thin enough to transmit light. This passage, he adds, can sometimes be observed in the same slice, and seen to be produced by the breaking up and progressive carbonisa-tion of the spores till they become entirely replaced by a uniform ground-mass.

Professor Williamson, on the other hand, concludes

from the examination of a suite of slices of Lancashire coals that this ground-mass is altered mother of coal, and he states that portions of the Better Bed, which under a low power appeared to be mineral charcoal, when ground very thin and submitted to a higher power, were reduced to the same condition as the dark ground-mass.

Professor Huxley states that all the coals he has examined agree more or less closely in their ultimate structure with the Better Bed; spores are always present, and in the purest and best coals they make up nearly the whole mass; and he accounts very satisfactorily for the preservation of this part only of the plants, on the ground that the resinous nature of the spores protected them from decay while the wood rotted away; the bark, which is rather less destructible, was the only part of the stem which escaped, and this appears in the mother of coal.

But though the Better Bed may be taken as a type of a large number of coals, there appear to be others, nearer home than Nova Scotia, into whose composition spores do not enter so largely. Professor Williamson found that the Oolitic coal of Cloughton Wyke, near Scarborough, contains no spores. He also noticed that in the coals of Worsley in Lancashire abundance of spores and excellent quality do not always go together. The Bin's Mine, which is rich in spores, is a worthless coal: spores were also found in crowds in some fireclays and ironstones. Mr. Binney, again, has furnished us with notes on the composition of some Scotch coals. He noticed that the bark of fossil Lepidodendra and Sigillariæ is often converted into coal without microscopic structure and resembling caking coal, and he believes that such coal consists very largely of the bark of these trees. Other coals, in which what he believes to be microspores were present in large numbers, burn with a brilliant flame and empyreumatic

odour like Boghead cannel, and he thinks that coals of
this class are largely made up of microspores, even though
it is now no longer possible to detect them. Splint coal,
on the other hand, full of macrospores, burns and smells
like ordinary hard coal.

All observations then show that coal is composed of
vegetable matter, which is the point we want to prove;
but, as far as our present knowledge goes, it seems likely
that the portions of the plants which furnished the
material for the coal are not in all cases the same.

That some coals are almost entirely composed of
spores there can be no doubt; but when we reflect on the
minute size of these bodies, the number required to
make up a seam of coal is almost enough to shake our
belief in this fact, if direct observation had not placed its
truth beyond question. Instances, however, of enormous
accumulation of similar substances in modern times are
not wanting. In Inverness-shire a great shower of the
pollen of the fir took place in 1858; the ground was
covered by a layer of this substance, in some places to a
depth of half an inch, and the deposit was noticed at
places thirty-three miles apart. The whole surface of the
great lakes in Canada is not unfrequently covered by a
thick scum of the same pollen. Similar occurrences have
been noticed in the forests of Norway and Lithuania.

Instances like these are not without their use, for they
enable us to picture to ourselves what must have been in
many places a constantly recurring incident during Carboniferous times; but there ought not to be any real difficulty
in accepting the view of the spore origin of coal. Even
microspores, small as they are, are larger than a *Globigerina*
shell, and the total thickness of all the coal seams in the
richest coal field falls far short of that of the chalk; but
no one raises any difficulty when chalk is said to be made

up of foraminiferous shells, and when its real meaning is apprehended, this is a far more startling statement than that spores can accumulate in sufficient quantity to form a seam of coal. Any little hesitation that may at first be felt will be considerably lessened if we bear in mind the size of Lepidodendron. Look at the cloud of spores that can be shaken out of a spike of the tiny club moss, then picture a well grown tree thickly hung with cones, every one of which discharges a no less copious shower, and then imagine a forest of such trees. Will not the yield be enough to satisfy even the most sceptical?

Those who are still slow to believe will perhaps be re-assured when they are told that in Australia there are two combustible minerals of much more recent date than our English coal beds, which are made up to a large extent of the very same kind of minute bodies as are visible in our older coals. One of these is called White Coal, and the other Tasmanite. Tasmanite is a shale containing from 26 to 30 per cent. of combustible matter; microscopic sections show in it numerous sacs, exactly resembling those which occur so abundantly in the Better Bed coal, and which are doubtless the spores of a lycopodiaceous plant. It is the presence of these spores that gives it its inflammable character. The bed of Tasmanite is from six to seven feet thick, and covers an area whose limits have not been ascertained, but which is known to be miles in extent. White Coal is very similar to Tasmanite.

That coal is mineralised vegetable matter we may now look upon as a point thoroughly established. We have yet to explain in what way the enormous accumulation of vegetable material required for the manufacture of coal seams were gathered together.

The two points which are at once the most striking and the most difficult to account for in the formation of beds of coal are these. First, their singular purity; by which is meant their freedom from any admixture of substances which may not have been derived from a vegetable source. All coals when burnt leave behind a residue called ash; so does wood; and in many cases the ash of coal is probably nothing more than that of the wood from which the coal has been derived. In other cases a part at least of the ash is clayey or sandy matter that has been mixed with the vegetable material. In the case of all good coals however the amount of ash is very small. The second point that attracts notice, is the constancy with which seams of coal maintain pretty much the same character and thickness over very large areas.

In order to account then satisfactorily for the formation of Coal, we must show by what means broad sheets of vegetable matter, nearly uniform in thickness and all but free from admixture of foreign substances, could have been spread out over areas hundreds of square miles in extent.

A very little reflection will show that there are two very obvious reasons why coal could not have been formed in the same way as the shales and sandstones among which it occurs, and that it is quite impossible that the plants which furnished the material of coal could have floated down rivers into seas and lakes, and been spread out in sheets on the bottom of the water. That great matted masses of plants and trees do gather together in rivers, and becoming water-logged sink to the bottom of the sea into which the river discharges itself, is an occurrence that happens in the case of many streams, the Mississippi for instance. But such a thing as a river that brings down nothing but drifted plants is utterly inconceivable. Sand and mud will always be transported as well, and however large and numerous the masses of floating vegetation may be, the deposit formed must be a mixed one, consisting partly of vegetable matter and partly of sediment.

The beds formed out of such a mixture will, no doubt, be more or less coaly or carbonaceous in character, but they will not possess the purity which is one of the distinguishing marks of true coal, and if an attempt be made to use them for fuel, they betray their origin by the large quantity of ash they leave on being burnt.

Again such floating masses of vegetation will settle down very irregularly, at spots more or less widely separated from one another, and the intervals will be filled up with sandy or clayey deposits all but free from vegetable matter. In the strata formed in this way there will be portions containing a large quantity of carbonaceous matter, but they will be scattered about in patches, and there will be nothing corresponding to the uniformity in thickness and character which coal seams exhibit in so striking a manner. Just such patches we do find not unfrequently in the sandstones of the Coal Measures, now converted into nests of coal, often bright and pure, but thinning away on all sides, so that no one would ever speak of them as seams. It may therefore be looked upon as a settled thing that coals has not been formed out of drifted vegetable matter.

There remains the explanation that the plants out of which coal was formed sprang up, grew, and died on the spot where the coal is now found.

Imagine a broad, flat, swampy expanse, thickly covered with a luxuriant vegetation, such a tract of country for instance as those pestilential, marshy jungles which are met with in Central Africa. One year's crop of plants arises, dies, and falls to the ground, and is succeeded in the next year by an equally luxuriant growth that in its turn withers and drops. Let this go on year after year for many a decade, perhaps even for centuries. The close of each season will see the surface strewn with dead vegetable matter; part will decay, other portions, such as bark and spores, are better able to resist decomposition and do not pass away; as year after year layer is added to layer, the accumulation grows in thickness, till at last a broad and deep stratum of the remains of fallen herbs and trees spreads over the whole area.

A deposit formed in this way will contain nothing but vegetable matter, it will also be fairly continuous and tend to vary but slightly in thickness, and so it will possess the two peculiarities which seams of coal exhibit in so marked a manner.

Here then we seem to see our way to a consistent explanation of the origin of coal, and the theory that coal was formed in this way is fully borne out by the following facts.

It was noticed by the late Sir W. Logan that every seam of coal rests upon a bed of rock of very peculiar character, which goes by the general name of ' seat-stone ' or ' seat-earth,' and is also called ' underclay,' ' warrant,' and in Yorkshire ' spavin.'

Seat-stones vary very much in their composition, the generality of them are clays, but they often contain a large admixture of sandy matter; sometimes they are almost wholly made up of silica, when they form a very hard flinty rock known as Ganister.

But whatever be their composition, all seat-stones agree in two points ; they are perfectly devoid of stratification, and they contain certain vegetable fossils called Stigmaria.

These Stigmaria are long, cylindrical, branching bodies, covered with little round pits arranged in a lozenge-

shaped pattern; each pit contains a small nipple, and to
the nipple there is attached a long ribbon-shaped black
filament. The Stigmaria lie horizontally in the underclay, and the filaments run out in all directions till the clay
is frequently one thickly matted mass of them. The
whole thing has a very rootlike look, the Stigmaria being
the larger branches of the root and the filaments the rootlets, so that when Brongniart discovered that the internal
structure of Stigmaria is very similar to that of a plant
called Sigillaria often found fossil in the Coal Measures,
and inferred from this that one was the root of the other,
geologists were not unwilling to accept his conclusion
though Stigmaria had never then been found actually
attached to Sigillaria. It was reserved for Mr. Binney
to supply this link in the evidence. He discovered in a
railway cutting near St. Helen's in Lancashire a number
of trunks of Sigillaria standing erect as they grew with
roots still attached to them. The roots were Stigmaria,
and the rock into which they struck down, which was of
course the soil on which the trees grew, was the seat-stone
of a thin seam of coal.

The proof was now complete. All coal seams rest on
old soils, and practically consist of nothing but vegetable
matter. The abundance of roots shows that the soils
supported a luxuriant growth of plants, and these, as they
died and fell to the ground, would supply exactly the
material for the manufacture of coal.

The herbs and trees then of which coal is formed
grew on the areas where the coal now occurs; the ground
on which they grew was probably such as we could hardly
call dry land, but was rather a spread of swamps and
marshes; in some cases it may have been covered with
water of a moderate depth, so that the roots and the
lower part of the stems were submerged, while the larger

part of the trunks and the branches rose into the
air.

But whether the plants grew in marshes or on ground
just submerged beneath water, coal may be looked upon
as a product of what was practically land growth, and

FIG. 2.

Erect Sigillaria rooted in an Underclay full of Stigmaria.

Sandstone .....

Shale ....

Sandstone ....

Coal ....

Underclay,
with Stigmaria.

each coal seam indicates the existence of a land surface
at the time it was formed.

One variety of coal, Cannel, requires separate notice,
because it has probably been formed in a somewhat different
manner to black or common coal.

Cannel coal does not soil the fingers, and the best
examples are compact enough to take a polish; it breaks
with what is called a conchoidal fracture, that is in
curved surfaces with concentric ridges like the pattern of
a ribbed shell; it lights readily and burns with a bright
flame. The distinctive point in the composition of cannel
is the large proportion of hydrogen which it contains; it
is this that makes cannel specially serviceable for making
gas.

Cannel coals always occur in dish-shaped patches thinning away to nothing on all sides; they frequently merge insensibly into highly carbonaceous black shale; and they contain occasionally the remains of fish.

The presence of fossil fish in cannels show that they must have been formed under water, and they probably consist of vegetable matter which was drifted down into ponds or lakes and lay soaking till it became reduced to a pulp. The deposit was of course limited in extent by the banks of the sheet of water in which it was formed, and hence the lenticular shape which beds of cannel exhibit. A certain amount of mud would of course be brought into the water along with the drifted plants, but being heavier than they it would fall down first carrying with it enough decomposing vegetable matter to stain it black; in a certain distance all the mud came to the bottom, and the vegetable residue floating on sank slowly and became spread out over the bed of the lake further on. Thus near the mouth of a river deposits of laminated carbonaceous mud were laid down, and these gradually contained less and less mud and more and more vegetable matter till they merged into a mass of vegetable pulp.

The maceration it has undergone has to a large extent effaced all traces of vegetable structure in cannel coal, but spores can now and then be still detected in it.

CHAPTER II.

THE GEOLOGY OF COAL—*continued.*

AN Egyptian monument thickly overspread with hierogly-
phic pictures, or a slab of stone covered with inscriptions
in arrow-headed characters may be objects of curiosity and
wonder; but till the chance discovery of a Rosetta Stone
or the persevering ingenuity of antiquaries enabled us to
assign a meaning to the symbols, they told us next to
nothing of the people who engraved them or of the events
they were intended to commemorate. Let the key however be hit upon, and a flood of light is at once thrown
on the manners of nations and the events of periods of
which no written history has come down to us.

Just so was it with the great monumental pile formed
by the rocks of the ground beneath us, on which nature
has inscribed symbolical representations of the events of
a long series of bygone ages. Its 'figured stones' were
eagerly sought after by collectors and earnestly pored
over by students; but none could fully read their meaning.
The less obvious lessons to be learned from the structure
and composition of the rocks themselves were hardly even
suspected. At last, after many false paths had been tried
and abandoned and some few partial successes had been
achieved, the right clue was found, and from that day
forward the task of deciphering the stony record has
been carried on with unabated energy and unvarying
success.

What the nature of that clue was, has been explained in the previous chapter. Let us in the present chapter apply the principles there laid down to the case of the Carboniferous rocks of the northern half of England and the Lowlands of Scotland, and see how far we can by the aid of those principles draw out a connected account of the condition of the country and the succession of events that went on during the period when those rocks were being formed.

The Carboniferous system of rocks can be subdivided into the groups shown in the table on p. 35. It will be seen from that table, that these groups differ from one another in being made up of different kinds of rocks; in one limestone is the main constituent, another is distinguished by the number and excellent character of its seams of coal, and so on. These differences in the character of their component rocks are quite sufficient to define the subdivisions, but underlying them there is a difference of far greater moment. Each kind of rock requires special physical conditions for its formation. By physical conditions are meant such things as depth of water, distance from shore, extent and shape of adjoining land, and such like. Limestone, it will be recollected, was formed in clear water far from the coast, conglomerates along the coast-line, coal grew on land. Well then, if we bear this in mind, we shall see that each subdivision corresponds to a period of time during which the physical conditions were different from those which prevailed during the periods corresponding to the other subdivisions. To borrow an illustration from history, if we had before us a collection of all the coins struck in France during the last hundred years, we should learn from them, that the French nation had been at one time under monarchical,

at another under republican, and at another under
imperial government. Not less unmistakeably does the
nature of the prevailing rock in each of the subdivisions
of the Carboniferous system point out that during part
of the Carboniferous epoch the country was submerged
beneath the clear waters of a central sea; during another
part was overspread by the muddy waters of a shallow
fresh-water lake or estuary; and was again from time to
time a broad, marshy fen. It will be my aim to explain
what were the physical conditions of the different portions
of the Carboniferous epoch, and how the passage from one
state to the state that followed was brought about; but
before we come to this, our main subject, it will be
necessary to make a few enquiries into the nature and
shape of the floor on which the lowest subdivision of the
Carboniferous rocks rests.

Over a large part of the area we are concerned with,
this floor is formed of a group of hard, slaty rocks, to
which geologists have given the name Silurian. These
beds may be seen underlying the Carboniferous Limestone
in the dales on the north side of Ingleborough, in the
Crummack Valley and in Ribblesdale near Clapham, in
Gordale, and around the outskirts of the Lake district.
Now, though the Carboniferous Limestone comes directly
on the top of the Silurian rocks at the points mentioned,
and probably over the whole of the intervening area, the
one was not formed immediately after the other, but a
very long period elapsed between the time when the
deposition of the Silurian rocks was completed, and the
day when the lowest bed of the Carboniferous Limestone
was laid down. We can say that this was the case,
because elsewhere a group of rocks known as the Old Red
Sandstone lies between the Silurians and the Carboniferous Limestone, and as this group reaches a thickness

Carboniferous Rocks of the North of England.

	Composed of	Formed in	
Upper Coal measures	Red sandstones and shales. Thin limestones. Few coals.	? Closed freshwater lake.	Fresh or brackish water.
Middle Coal measures . .	Shales. Few sandstones. Many thick and good coals.	Shales and sandstones in freshwater lake or estuary, now and then flooded by sea water. Coals on land.	
Lower Coal measures . . .	Many sandstones. Shales. Few thin and poor coals.		
Millstone Grit	Many thick coarse sandstones. Shales. Very few thin and poor coals.		
Yoredale Rocks	Shales, with thin impure limestones.	Shallow muddy sea.	
Carboniferous Limestone .	In Centre. Solid pure limestone. On North and South. Shales, sandstones, and limestones.	Clear water of the centre of the sea. Shallow muddy water of the margin of the sea.	Salt water.

Uneven floor of Silurian Rocks.

of several thousand feet, a long time must have been occupied in its formation.

What was going on during this time over the area we are dealing with? Because there is there no Old Red Sandstone, we conclude that the Silurian rocks were raised above water, and that during the whole of the Old Red Sandstone period, this district was dry land. The Silurians were uplifted by being crumpled up into arches or folds, so that their beds are no longer in the horizontal position in which they were laid down, but stand up at high angles. As the rocks rose, the crests of the arches were planed away by the sea, and the land, when it first emerged from beneath the waves, had an even surface. But no sooner was a tract of dry land established, than another set of wearing agents came into play and worked their will upon it. Rain, frost, and running water ate deeply into the flat surface, and carved it out into hills and valleys. The floor which was destined to support afterwards the Carboniferous Limestone was thus very uneven, rising here into peaks and ridges, and sinking there into hollows and valleys.

It was not till after this process had been completed, that the formation of the Carboniferous rocks began.

The lowest member of this group, the Carboniferous Limestone, is a limestone crowded with marine fossils, and therefore formed beneath the waters of the sea. This district then which had remained above water for so long a period was now lowered till it was sunk beneath the sea level and became covered by salt water, and the bottom of the sea so produced was very uneven. The irregularity in the ocean-bed would cause the Carboniferous Limestone to vary very much in thickness; where a deep depression existed, bed after bed would be piled one on the other, till the hollow was filled up, and the thickness of the deposit would be very large; when there was a ridge or elevated plateau, a smaller thickness would accumulate. Now this is just what we find to be the case. In Derbyshire the limestone is so enormously thick that we never see the bottom of it; there are certainly more than 2,000 feet, perhaps more than 4,000 feet of it. Under Ingleborough there is not more than 600 feet of limestone; near Clitheroe it is at least 3,250 feet thick. If we go as far south as Leicestershire, this' subdivision does not, on the most liberal estimate, reach a thickness of more than two or three hundred feet.

What has been just said serves very well to show how facts, which at first sight seem to have very little connection, are really closely related. The absence of the Old Red Sandstone in the north of England enables us, when all its bearings are taken into account, to explain satisfactorily the striking irregularity in thickness of the Carboniferous Limestone of that area.

The depression which produced the Carboniferous Limestone sea did not come to an end when that sea was established. As we go along, we shall find abundant proof that the sea bed and the surrounding land were steadily going down during the whole of the Carboniferous period. The subsidence went on at a slow but variable rate, and every now and then stopped altogether for a time.

Our next step will be to mark out the boundaries of the sea in which the Carboniferous Limestone was formed. The principles which must guide us in this investigation have been already explained, and the result of their application is the map in fig. 3.

It would be wearisome to go into the details of the evidence for the position of every point on the coast-line; indeed, there are portions of that line so very conjectural

that perhaps it will be only prudent to observe a judicious
reticence respecting them. One or two of the more
salient points in the method employed in the construction
of the map must however be noticed. The northern and
southern boundaries of the sea are fixed by the following

FIG. 3.

*Map showing the general distribution of land and water during the forma-
tion of the Carboniferous Limestone of Great Britain and Ireland.*

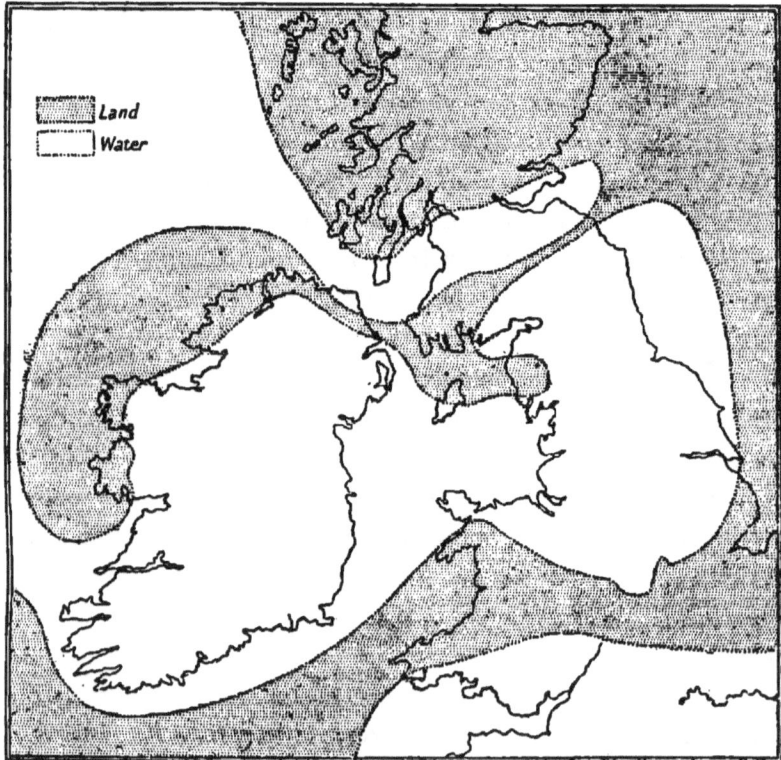

considerations. The Carboniferous Limestone of Derbyshire is from top to bottom a succession of beds of very
pure limestone, many of which can even now be seen to be
wholly made up of the hard parts of marine animals.
There are occasionally thin partings or wayboards of clay

or shale between the limestone beds, but they are few in number and very thin. I once measured a section which showed more than 1,500 feet of limestone, and if all the clay bands which occur in that section were laid one on the top of another, they would not make a bed of more than a few yards thick. The Carboniferous Limestone maintains the character throughout Lancashire and Yorkshire.

Derbyshire, Yorkshire, and Lancashire therefore were in early Carboniferous times covered by portions of the sea, so far removed from land that no sediment found its way out to them.

When we reach Cumberland, the Carboniferous Limestone undergoes an important change. It still contains many beds of pure solid limestone, but interstratified with these are sandstones and shales, often of considerable thickness.

In Northumberland and on the Borders, the sandstones and shales have increased so much in bulk that the limestone becomes quite a subordinate element. It occurs in a few thin beds buried in an enormous body of shale and sandstone, and forms so insignificant a member of the whole mass that it might easily be overlooked; indeed the application of the name Carboniferous Limestone to the group is apt to seem to the uninitiated a striking instance of the perversity with which geologists cling to the *lucus a non lucendo* principle in their nomenclature.

In a word as we follow the Carboniferous Limestone from Derbyshire to the borders of Scotland we find sandstones and shales coming in and gradually thickening northwards till they all but replace the limestone completely.

Following the principle laid down in the previous chapter, we conclude that land lay not far to the north of the Scotch Border. The coast line of those days probably ran somewhere about the south-eastern flank of the line of hills which stretch from St. Abb's Head to Galloway and are known as the Southern Uplands of Scotland.

South of the Derbyshire exposure of the Carboniferous Limestone that rock is for some distance hidden beneath a tract of New Red Sandstone. It again comes to the surface in Leicestershire. Here it is thin and frequently earthy, and it contains a great deal of interbedded shale. In this direction then we are again drawing near a coast line. In South Staffordshire the Coal Measures rest directly on the Silurian rocks and there is no Carboniferous Limestone at all. South Staffordshire therefore was dry land during Carboniferous Limestone times. A southern shore line then of the sea ran somewhere between Leicestershire and South Staffordshire.

Thus we get approximately the position of a northern and southern coast.

The following evidence enables us to carry on the southern shore-line to the west. The Carboniferous Limestone of Derbyshire and Flintshire is thick and pure, but in Anglesea the rock contains interbedded sandstone and shale. Now we know that the Lower Silurian rocks of Caernarvonshire were raised into dry land at the end of the Lower Silurian epoch; it is likely enough that they continued above water for a long period; that they formed a portion of the southern boundary of the sea we are dealing with; and that from them were derived the materials for the sedimentary beds intercalated among the Carboniferous Limestone of Anglesea.

To the north-east we have no data for fixing the exact extent of the water. The following considerations however lead to the conclusion that land was not very far distant in that quarter. A great many of the gritstones of the Carboniferous group are obviously nothing but granite, or some crystalline rock allied to granite, pounded up and reconsolidated, and in some cases the pounding has been so imperfectly carried out that the original minerals still exist in pieces large enough to allow of any peculiarities which they possess being recognised.

Mr. Sorby has gone minutely into the question, and he finds that the grains of quartz in these grits are much more like the quartz of the granites of Norway and Sweden than that of any other granite he is acquainted with. The felspar also, when unchanged fragments are met with, bears the same resemblance to that of the Scandinavian granites.[1] But Mr. Sorby also finds that the majority of the quartz grains are as angular as if they had never been subjected to attrition; they cannot therefore have travelled very far from the parent rock, certainly not over anything like the distance which now separates Scandinavia from England.

But there is nothing improbable in the supposition that the Scandinavian peninsula once extended much further to the south-west than now, and that its prolongation was composed of the same kind of rocks as make up the bulk of it at present.

Such a supposition exactly meets the requirements of the case. It will give us the right kind of rocks to furnish the requisite materials, and the rocks will be near enough to allow of these materials reaching their destination in the requisite state of coarseness. We conclude therefore that to the north-east the sea was bounded by land formed by a south-westerly prolongation of Scandinavia.

[1] *Monthly Microscopical Journal*, Anniversary Address, 1877, pp. 20, 21.

We will now see what information can be obtained in Ireland as to the boundaries of the sea we are mapping out. Over the larger part of the sister isle the Carboniferous Limestone consists of three members. The top and bottom divisions are made up mainly of clean organic limestone; the middle division, called the Calp, is more earthy. The upper and lower portions then of most of the Irish Carboniferous Limestone were formed in clear water remote from the shore. The conditions that prevailed during the formation of the Calp were probably these. While the bottom limestone was growing up, the sea bed was steadily sinking; then the subsidence ceased, and, as the limestone gradually grew in thickness, it filled up the sea, and the centre of Ireland became covered with water of only moderate depth. The rivers now, instead of being checked at a small distance from the coast, would retain a considerable velocity far out to sea, and would spread over the whole area the sand and mud they brought down; the calcareous matter furnished by animals would be mixed with sediment, and earthy limestone would be the result. After a while subsidence began again, tracts of deep and clear water were reestablished, and in these the upper clean limestone accumulated.

Such is the nature of the Carboniferous Limestone over the larger part of Ireland. In certain quarters however it puts on characters analogous to those which it wears on the Scotch Borders and in Leicestershire, and these enable us to lay down the coast lines of the sea in which it was formed.

Towards the north and north-west the rock becomes interleaved with shale and sandstone, and remains of land-plants and thin coals occur in the Calp. Here is evidence that in this quarter we are in the neighbourhood of a coast line. The highlands of Connemara, of the extreme west of Mayo, and of Donegal are probably the remains of the land that bounded the Carboniferous Limestone sea on this side. The rocks of these districts are highly metamorphosed Lower Silurian beds; they were upheaved at the close of the Lower Silurian period, and it is likely that, with the exception of a few unimportant submergences, they have remained dry land ever since. A shore line then wrapped round what is now the north-west coast of Ireland, and it is likely enough that it was continuous with the shore line that ran along the southern flank of the Southern Uplands of Scotland.

Again, in the south of Ireland the Carboniferous Limestone is from 2,000 to 3,000 feet thick and pure about Kilkenny, but when followed to the south-west through Cos. Waterford and Cork it thins away entirely and is replaced by shale. Still further to the south-west in Co. Kerry great masses of coarse sandstone, known as the Coomhoola Grits, come in among the shale. It would scarcely be possible to find a better instance than this of a sheet of rock which at one spot is thick and pure limestone, then merges into shale, and then gradually grows sandy till it consists in large measure of gritstone. The conclusion is obvious: in travelling from what is now Kilkenny towards Kerry we should in Carboniferous Limestone times have passed from a sea of clear water towards land. The shore line here was probably a continuation of that which ran through Caernarvonshire.

By some such considerations as these the boundaries of the Carboniferous Limestone sea can be approximately fixed on the north, east, and south. It was certainly land-locked on these three sides; it must have had some communication with the ocean, and the west is the only quarter where an opening is possible. There was therefore probably a strait or straits on the west connecting this inland body of salt water with the open ocean beyond. We have been able to show that the northern and southern coast lines approached one another on the west so closely that these outlets cannot have been of any great breadth.

In this mediterranean sea the formation of the Carboniferous Limestone went on in the manner already described. In the central portions, where the water was clear, animals built up vast masses of pure limestone, the thickness of the deposit being measured by thousands of feet wherever there was a deep depression in the sea bed, and not exceeding a few hundred feet where the bottom rose into ridges or platforms. Nearer the shore, where the water was continually receiving muddy and sandy sediment, the deposits were of a mixed character, clayey, sandy, and impure calcareous beds alternating with one another.

And now let us pause for a moment, and reflect on the change in physical conditions which the accumulation of the great mass of the Carboniferous Limestone would give rise to. All the deeper parts of the water would be gradually filled up, and the sea would at last become over its whole extent as shallow as it had been at first only along its edges. When this change had been brought about, the kind of deposit, that had been at the beginning confined to the margin, would extend itself outwards, till at last it spread over the whole length and breadth of the water.

We may expect then that the Carboniferous Limestone would be overlaid by deposits, in which sandy and muddy beds would play an important part, and this is just what we find to be the case. The Yoredale rocks, which come next in ascending order, resemble very closely those portions of the Carboniferous Limestone group which were deposited near the shore. Shale, with occasional beds of sandstone, forms the bulk of the subdivision, but limestone is always present, frequently earthy and impure, and sometimes little better than calcareous shale.

In fact during the deposition of the Yoredale rocks there was a repetition, over a wider area and under a more pronounced form, of the conditions that had existed locally during the formation of the Calp limestone of Ireland.

Really and truly then, though we can over a large part of our area subdivide the lower part of the Carboni-ferous rocks into a Carboniferous Limestone and a Yoredale group, there are other parts in which the correspond-ing rocks from top to bottom might be called Yoredale rocks.

Wherever there was shallow muddy water, beds cor-responding to the Yoredale type were laid down; wherever an area of clear water existed, its deposits took the form of Carboniferous Limestone.

But though the Yoredale rocks were formed in water of no great depth, they reach in the aggregate a very considerable thickness, at some spots not less than several thousand feet. This apparent contradiction is easily got over. We have only to suppose that during the formation of this group the sea bottom was steadily sinking, and that the space through which it went down in any time was about the same as the thickness of the sediment de-posited on it during the same time—in other words, that the rate at which the sinking of the bottom tended to deepen the water, and the rate at which deposition tended to fill it up, were nearly equal. With such an adjustment deposits of any thickness might accumulate, and the water still remain shallow all along.

We have already promised to bring proof that a slow downward movement of the land was going on during the whole of the Carboniferous period, and we have here evidence that it was in progress during Yoredale times.

With the exception of a few drifted land-plants, the fossils of the Yoredale rocks are all marine, and are all found in the Carboniferous Limestone. They are in fact the remains of those members of the Carboniferous Limestone fauna which were sufficiently indifferent to their surroundings to be able to live on under the changed conditions of Yoredale times.

The rocks that were being deposited over what are now the Lowlands of Scotland while the Carboniferous Limestone and Yoredale rocks were accumulating in the north of England, differ in so many respects from their English equivalents that they require separate consideration.

The Scotch Carboniferous rocks occupy a strip of comparatively low lying land, which stretches across the country from the Firth of Forth to the Firth of Clyde, and is bounded on the south by the Southern Uplands, and on the north by the Grampians.

The lower portion of these beds falls naturally into two divisions, which go by the name of the Calciferous Sandstones and the Carboniferous Limestone.

The Calciferous Sandstones, in their turn, admit of two subdivisions—a lower, consisting of sandstones, many of which are coarse and conglomeratic,[1] and an upper, made up of a very irregular assemblage of sandstones, shales, limestones, and coals; this upper portion is sometimes distinguished as the Cement-stone group. Both divisions contain large masses of rock of volcanic origin.

The great mass of grit and conglomerates which makes up the base of the Calciferous Sandstone group was evidently deposited in shallow water and at no great distance from land. While these beds attain a great

---

[1] These beds are sometimes called Upper Old Red Sandstone.

thickness at some spots, they are altogether wanting at others. In some cases this is no doubt owing to the fact that the coarse materials were thrown down in banks; in other cases the sea bed may have been traversed by ridges and shoals, and the hollows between these were filled up by accumulations of sand and pebbles. The typical character of these beds comes out with singular distinctness in the island of Arran, where they consist of thick masses of very coarse quartzose conglomerate alternating with more finely grained red sandstones. A large number of the pebbles in the conglomerates are very imperfectly rounded, and possess edges so sharp and jagged that it is evident they have not travelled very far. We are clearly here dealing with piles of shingle heaped up in close proximity to the shore.

If we suppose Scotland to sink to the extent of a few hundred feet, we should obtain exactly the kind of sea suited to the formation of this group of rocks. The Firths of Forth and Clyde were united by a shallow strip of water. The land rose sharply on both the north and south sides into hills formed of hard quartzose rocks and traversed by numerous veins of white quartz. Down the slopes great piles of gritty and pebbly debris were hurried into the water and heaped up in irregularly shaped, moundy masses. As the land slowly sank, the deposit was gradually added to till it grew to a considerable thickness.

There is evidence then to show that at the beginning of the Carboniferous period a narrow strip of sea ran along what are now the Lowlands of Scotland. Are there indications of any tract of water still further north? There are not; on the contrary there is every reason to believe that the whole of the Highlands was dry land at this time; and most likely this land surface stretched away

across what is now the North Sea up to Norway and
Sweden, and was continuous with the land which bounded
on the north-east the Carboniferous Limestone sea of the
north of England. The Scotch Carboniferous rocks were
therefore formed in a long, narrow inlet, which opened
out to the west and ran up in a north-easterly direction
into the heart of a large tract of land, formed very largely
of hard, crystalline rocks.

We now come to the Cement-stone group, the upper
division of the Calciferous Sandstones. It is scarcely
possible to give any adequate description of the irregularity of these beds. As they are traced across the
country the different members replace one another in
every conceivable way. Limestones and shales occur in
patches, surrounded on all sides by sandstones. Coals
pass away into black band ironstone, or shade off into
black shale, so that no two sections of the measures taken
a short distance apart show anything approaching agreement in their details.

These beds must have been deposited in a group of
lagoons or creeks with muddy shoals and banks between
them. In every pond-like patch of water shales and limestone were deposited, while the intervening low, swampy
banks supported vegetation, out of which coal was formed.
Such a state of things would be produced if we suppose
the gradual subsidence of the land to be suspended or to
go on slowly, while a supply of coarse sediment was poured
into the water without intermission. Sandbanks would
be everywhere established, their summits would gradually
rise into swamps, and the spaces between the banks would
become the site of pools.

The Carboniferous Limestone group resembles very
closely the rocks that bear the same name in the
extreme north of England, and is, if possible, less entitled

to the name even than they. Limestone enters into its composition to the extent only of a few thin beds; sandstones and shales are the prevailing rocks, and the subdivision also contains coals, some of which are thick and pure enough to allow of their being worked.

In these beds we have evidence of a time of incessant oscillation. At one time a somewhat rapid subsidence laid the whole area under salt water in which bands of marine limestone grew up. The neighbourhood of land however never allowed the water to become clear, and the calcareous matter furnished by animal agency was always more or less mixed with sediment. At other times the rate of sinking was checked, and deposition shallowed the water till it became so fouled with mud and sand that limestone-forming animals could no longer abide it. Now and again the sinking ceased altogether for a while, the water was filled up, and swampy tracts formed, on which vegetable growth gave rise to beds of coal.

There was no doubt a close connection between these changing circumstances and the volcanic action which was busily at work in the south of Scotland during early Carboniferous times. Volcanic areas are notorious for their instability; the land is frequently in a state or incessant oscillation; now up, now down, and now for a while stationary. The north of England, though not altogether free from volcanic outbursts at this date, was not so crowded with volcanoes as the Lowlands of Scotland, and the course of events there had a more even tenor.

We have now narrated the history of the periods during which the Carboniferous Limestone and the Yoredale rocks were in course of formation, and it has appeared that the transition from the condition of Carboniferous Limestone times to those of the Yoredale period was brought about by no great physical revolution, but was simply the natural

E

consequence of the deposition of the Carboniferous Limestone itself. The two epochs in fact stand to one another in
the same relation as two scenes in the same act of a play.

But it is otherwise when we pass on to the higher subdivisions of the Carboniferous system. These rocks furnish
proof that the physical conditions of our area underwent,
in one respect, a very material change at the end of the
Yoredale period, and that with them we are entering on a
new *act* in the drama. The Millstone Grit and the Lower and
Middle Coal Measures have, as the table on p. 35 shows,
each of them certain minor peculiarities in lithological composition which distinguish them from one another, but in
broad general characters they agree so closely and differ
so widely from the two subdivisions below them, that it is
impossible to resist the conclusion that they were all three
formed under substantially the same conditions, and that
these conditions were altogether different from those which
gave rise to the Carboniferous Limestone and the Yoredale
rocks.

These three subdivisions consist of sandstones and
shales with coal seams ; limestone is so excessively rare in
them that practically it may be said to be absent altogether. Their most abundant fossils are land-plants,
drifted or embedded on the spot where they grew. The
animal remains are represented in certain beds by an
abundance of individuals, but the species are not
numerous. Their bearing on the question we are now
handling is a matter that requires careful consideration.

We will take first the Mollusca. These may be divided
into two groups which, as a rule, keep quite apart from
one another, and which there is no satisfactory evidence
to show have ever been found associated together in the
same bed.[1]

[1] I know very well that in making this positive assertion I lay myself

The first group includes shells of the genera Anthracosia, Anthracomya, and Anthracoptera. Some palæontologists have unhesitatingly pronounced these to be freshwater forms, but the verdict of naturalists is not unanimous on this point; and indeed it seems very questionable whether the relationship of these shells to recent forms can be decided with sufficient certainty to allow of the formation, by this method, of a positive opinion as to their habitat. The question, if it is to be decided at all, must be settled by collateral evidence.

With these shells there are often associated Entomostracous Crustaceans, some of which have been referred to a recent fresh-water genus Candona, and an annelid *Spirorbis Carbonarius.*

The mollusca of the second group are all unquestionably marine and some of them Carboniferous Limestone species. These marine forms are found only in a few thin bands separated by wide intervals.

These two faunas recur, the first frequently, the second not so often, throughout the whole thickness of rocks from the bottom of the Millstone Grit to the top of the Middle Coal Measures; they were therefore contemporaneous. But they are rarely, if ever, found intermingled. One we know was marine. Is it not, to say the least, likely that the reason why they keep so markedly apart is that the other was fresh-water?

If this be admitted, the portion of the Carboniferous rocks above the Yoredale group must have been deposited

open to a charge of presuming to differ from so eminent an authority as Mr. Binney, who has stated that the two groups do occasionally occur together. It is just possible that even so keen-eyed an observer may not have been always quite careful enough to fix the exact *gisement* of his specimens, and may have quoted fossils from the same bed which came from two closely adjoining layers. But even if this is not the case, I feel sure that the association of the two groups is the exception and not the rule.

in the main in fresh, or possibly in estuarine, waters; and the Carboniferous Limestone sea must at the close of the Yoredale period have been converted into a fresh-water lake or an estuary : now and again however the sea burst in and overflowed the area, and during its incursions bands containing marine fossils were laid down.

During Carboniferous Limestone times, the inland sea and the outer ocean were inhabited by very much the same assemblage of creatures; but the change in the inland sea from salt to fresh water drove the marine animals out of it into the open ocean beyond, where some species lived on unchanged, and others in lapse of time became modified. At the same time a fresh-water fauna, which had hitherto dwelt in the rivers, established itself in the lake. There were therefore living side by side two distinct faunas ; one, consisting of Anthracosia and its companions, inhabited the lake ; the other dwelt in the ocean outside. As long as the salt water was kept back, the deposits formed enclosed only shells belonging to the Anthracosia group; every time it forced its way in, it drove back these creatures up into the rivers, and brought with it marine shells, some of which were identical with forms that had lived in the inland sea while the Carboniferous Limestone was being deposited, and others were the modified descendants of the Carboniferous Limestone species.[1]

We may now see what the other animal remains of the upper part of the Carboniferous system have to say to this view. The fish fall into two groups ; the Ganoids, represented now-a-days by the sturgeon and bony pike, and the Elasmobranchs, which include rays and sharks.

All the modern Ganoids, with the exception of the

[1] See Salter on the marine forms of the Coal measures, ' The Geology of the Country around Oldham.' *Memoirs of the Geological Survey of England,* p. 65.

sturgeon, are fresh-water fish; but the class was in bygone times so much larger than at present that it is perhaps hardly fair to infer from the habits of its scanty representatives in the modern day that Ganoids have been always confined to fresh water.   Sharks are as a rule marine, but as will be shown hereafter instances are known where they have been able to accommodate themselves to a fresh-water habitat.   The fish then give no help to the settlement of the question.

There remain the Labyrinthodonts; and as no marine amphibian is known, we may fairly conclude that these were fluviatile in their habit.

There is therefore a strong balance of evidence in favour of the view that the Millstone Grit and Lower and Middle Coal Measures are of fresh-water or estuarine origin.

The change from a land-locked sea with a narrow outlet, such as that in which the Carboniferous Limestone was probably deposited, to a fresh-water lake would be easily effected. If the outlet became nearly blocked up and the point of discharge was slightly raised above the sea-level, the salt water would be kept out, and the influx of river-water would freshen the contents of the basin. The marine animals would be killed or driven away, and fresh-water creatures would descend the rivers and take their place.   This has happened to some extent in the case of the Baltic.   We know from the contents of the old shell-beds along its margin that oysters and cockles once lived in it; but its waters have been gradually rendered so brackish by the large rivers discharging into it that these shells have either died off or are represented by dwarfed individuals.

The outlet may have been closed by the upheaval of a strip of the sea bottom, but it is quite as likely that it

became choked by the piling up of bars and sandbanks at
its mouth.

In either case the incursions of the sea admit of easy
explanation. If the barrier were permanent, we can easily
imagine that during the subsidence which was continuously
going on, a sudden drop might lower its summit beneath
the sea level, and that after a time it was restored by a
slight upward movement. On the other hypothesis, a sand-
bank would always be liable to be from time to time
breached during storms and subsequently repaired.

But whatever was its nature, it seems likely that the
barrier became more firmly established as time went on,
for the marine bands appear to be fewer in the Middle
than the Lower Coal Measures, and none have been detected
in the Upper Coal Measures.

We may suppose then that, without any change being
effected in its outline beyond the gradual increase in size
caused by the continual subsidence, our land-locked sea
became converted into a fresh-water lake.

The great mass of land lay to the north and north-east,
and was very largely composed of granite or other similar
crystalline rocks, which as they disintegrated under the
action of the air furnished the material for the sedimentary portion of the deposits we are now considering.
The quartz yielded the sand for the grits, and the quartz
veins the pebbles of the conglomerates, while the decomposed felspar supplied clay for the shales.

Here and there the shores may have been formed by
granite cliffs; and then we should have a state of things
such as Cameron describes in Lake Tanganyika, where
landslips are constantly launching masses of rock into the
water, to be there ground down and carried off into its
depths. Elsewhere there might be conditions similar to
those now existing along the western coast of the Gulf of

Genoa, where the mountains rise sharply from the shore, and the streams, when in flood, hurry down great masses of sand and shingle. Where the coast-line was less abrupt, rain and rivers would convey the products of surface decomposition from greater distances. As a rule it would seem that the source of supply was not far distant, for Mr. Sorby finds that nearly all the grains of quartz are angular. A few only are much worn ; and they may have been derived from dunes of blown sand.[1]

As the debris was swept down, it became arranged over the floor of the lake in the manner described in the previous chapter. The advent of a large supply of coarse sand would give rise at first to a line of sandbanks along the coast, just covered at top by a layer of shallow water ; across this belt of shallow water more sand would be rolled along the top of the bank and toppled over its face, thus forming a second bank in advance of the first. Thus the face of the deposit would be gradually pushed forward, and by the successive addition of bank after bank a broad sheet of sandy sediment would be spread out. As the bottom sank, another sheet would be laid on the top of this, and by a repetition of this process a thick accumulation of coarse sandy matter would grow up. Each river would furnish its quota, and a number of broad and thick masses of sandy sediment would be piled up over the bottom. The light, clayey matter which had been meanwhile float-ing in the water would slowly settle down, and fill up the spaces between the sheets of sand. Thus would be formed a deposit which, when consolidated, would present beds of sandstone wedging away, and replaced laterally by shale. At other times long ridges of sand may have been formed stretching away from the mouth of each river in the direction of its flow, till the floor of the lake was covered

[1] *Monthly Microscopical Journal*, 1877, Anniversary Address, p. 20.

by a great network of interlacing sandbanks. These were afterwards broken down and rolled to and fro by currents till their materials became spread out in a sheet of approximately uniform thickness. Thus we may account for the existence of a gritstone like the Rough Rock, which is remarkable for the persistency with which it maintains its thickness and coarse nature over very large areas. At other times when a somewhat more rapid subsidence deepened the water to such an extent that coarse sand could not be carried far out, the water of the lake would become charged with mud or fine sand throughout, and a thick and wide-spread accumulation of shale or finely grained, laminated sandstone would result. The most remarkable instance of this class of deposit in the area we are dealing with is the well known Elland Flagstone, which extends almost without break over the coalfields of Lancashire, Yorkshire, and Derbyshire, and is composed throughout the whole of its range of regularly bedded and closely grained sandstone of fine texture.

By all these changing circumstances, masses of rock resembling in every respect the sedimentary portion of the Millstone Grit and Coal Measures, would continue to accumulate as long as the sinking of the bottom went on.

But the downward movement was not without interruption; every now and then a pause occurred, and whenever this happened the water would tend to be filled up. Sandbanks formed shoals, which by degrees grew into islands; the channels between the islands were gradually choked up with sediment, till at last the whole or portions of the lake became replaced by swampy plains or morasses, across which sluggish rivers slowly wound their way. Vegetation spread from the adjoining land over the marshy flats, and under the moist and otherwise favourable conditions grew apace, till the whole became converted into a rank and tangled jungle.

How the fall year after year of trees and plants at the close of each season furnished the materials for a bed of coal, has been already explained.

After a time subsidence again set in; the sheet of fallen vegetation was lowered and overspread with water, and deposits of sand or mud piled on the top of it, which sealed it up and preserved it to appear in later days as a bed of coal.

The water came on so gently that usually the loose pulpy mass was not disturbed. Sometimes the influx was so gradual that the trees were not even overturned; their trunks seem then to have rotted away and broken off just at the surface of the water, and the submerged portions became buried in sediment, so that we now find them enclosed in shale and sandstone, standing as they grew with the roots still attached. Sometimes the roof of a coal is formed by a bed of shale containing marine fossils. In such cases the land perhaps went down with a sudden drop, which allowed the sea to overtop the barrier that had held it back, and inundate the flat. But the incursion was of comparatively short duration, for the marine bands are all of small thickness.

The broad, swampy, tree-covered flats on which coal accumulated are not without parallel at the present day. The Great Dismal Swamp of Virginia must present a very close miniature resemblance to them. Lyell [1] describes this great morass as having the appearance of a broad, inundated river-plain, covered with all kinds of aquatic trees and shrubs, the soil being as black as in a peat-bog. It is one enormous quagmire, soft and muddy except where

[1] *Travels in North America*, vol. i. p. 143.

the surface is rendered partially firm by a covering of
vegetables and their matted roots.   On the western margin
the land rises slightly above the surface of the swamp, but
the liquid mire of which the bog consists is not brought
down by streams rising in this quarter; it is usually
entirely formed of vegetable matters, without any admixture
of earthy particles.   On the north, east, and south, the
surface of the bog rises above the level of the surrounding
country, and the water coming down from the west flows
off in these directions.

Dr. Russell [1] gives a graphic description of the view
he obtained of the 'Great Dismal,' as he traversed it by
rail.   From the black waters there rose a thick growth
and upshooting of black stems of dead trees mingled with
the trunks and branches of others still living, and throwing out a most luxuriant vegetation.   The trees were
draped with long creepers and shrouds of Spanish moss,
which fell from branch to branch smothering the leaves in
their clammy embrace, or waving in pendulous folds in
the air.   Cypress, live oak, the dogwood, and pine,
struggled for life with the water, and about their stems
floated balks of timber on which lay tortoises, turtles, and
enormous frogs.   Once a dark body of greater size plunged
into a current which marked the course of a river; it was
an alligator, many of which come up into the swamp at
times.

The flats of the Coal Measures were probably less
completely covered with water than the morass just
described, but if we substitute Lepidodendra, Sigillariæ,
and Calamites for the modern vegetation, and put Labyrinthodonts in the place of alligators, the analogy between
the two will be very close.

Again there are enormous swampy expanses in the

___
[1] *My Diary North and South*, vol. i. p. 127.

delta of the Ganges and Brahmapootra known as the
Sunderbunds and Jheels, which serve to give us some idea
what the marshes of the Coal Measure period were like.
Dr. Hooker describes an immense area, drained by the
Soormah, a feeder of the Brahmapootra, which is scarcely
raised above the sea level and covers 10,000 square miles.
A great portion of this flat is covered by a dense growth
of tropical grasses, with low bushes here and there of
rattan-cane, laurel, and fig. A depression of ten to fifteen
feet would lay this and adjoining similar tracts under
water, and the Ganges, Brahmapootra, and Soormah would
then cover them with beds of silt and sand.[1]

It follows then that the Millstone Grit and the Lower
and Middle Coal measures must be looked upon as one
great deposit, composed from top to bottom of similar
alternations of rocks; characterised by fossils, which, if
they are not identically the same throughout, show merely
such specific modification as lapse of time might be reasonably expected to produce; and formed under substantially
the same physical conditions. We may now enquire on
what grounds this mass of strata, which from a broad
point of view constitutes only a single group, has been
parcelled out into three subdivisions. The reasons which
justify this separation rest on certain minor differences in
lithological composition which imply corresponding minor
changes in physical surroundings; and these differences are
as follows.

In the Millstone Grit the coals are very few, very thin,
very poor, and more or less local; in the Lower Coal
Measures the coals are still for the most part thin and of
second rate quality, but they are a decided improvement
on those of the Millstone Grit; the Middle Coal Measures
are distinguished by the number, thickness, and excellent

---

[1] *Himalayan Journals*, vol. ii. pp. 262-265.

quality of their coals, and also for the persistency of their coal seams over large areas.

These differences mean no more than this. During Millstone Grit times the pauses in the subsidence were few and far between, and when they did occur, did not last long; during the formation of the Lower Coal Measures pauses occurred rather oftener, and were of somewhat longer duration. The dirty quality of the coals of both subdivisions indicates that the vegetation accumulated on tracts which were either permanently submerged to a slight depth or were frequently flooded by muddy water, rather than on well established areas of marshy land. During the Middle Coal Measure period, on the other hand, pauses recurred frequently and lasted long; and during each pause the filling up of the water was so complete that the land surfaces formed were placed beyond the reach of mud-bearing inundations.

Again when we turn to the sandstones we find good reasons for the threefold partition. The Millstone Grit contains many thick beds of coarse massive gritstone and conglomerate, and for the class of rocks to which they belong these are remarkably persistent over large areas. In the Lower Coal Measures sandstones still play an important part; but, as a rule, they are neither so coarse nor strong as those of the Millstone Grit, and they vary much more from place to place both in thickness and texture. Throughout a large portion of the Middle Coal Measures thick sandstone beds are conspicuous by their absence, and when such beds do occur, they are with very few exceptions fine in grain; they are also markedly local, and can seldom be traced continuously for more than a few miles.

1 think we can see our way to an explanation of these differences, if we take into account the effect which pro-

longed subsidence would have on the surrounding land.
In the early part of Carboniferous times this land included mountainous tracts rising high into those regions
where rain, frost, running water, the violent beating of
storms, and all the army of subaerial denuding agencies
ply their destructive influence most incessantly and with
the most conspicuous results.   The consequence was the
formation of an abundant supply of coarse, sandy debris.
At the same time, owing to the configuration of the land,
the rivers would have a steep fall and would be competent
to sweep along heavy material.   During this period then
the materials for the formation of gritstone and conglomerate would be manufactured and transported in great
plenty.

But as time went on two causes were at work tending
to lessen the elevation and decrease the slope of the surrounding land.   The gradual subsidence and the cease-
less wear and tear of atmospheric denudation gradually
lowered the elevated tracts, so that they were acted on
less vigorously by subaerial denudation; at the same
time the rivers, descending by gentler gradients, lost by
degrees the power of moving coarse, heavy detritus.

So with the lapse of years the amount of sandy
sediment gradually grew less and less, and sandstones
formed a gradually decreasing item in the deposits in
process of formation.

The more rapid fall of the rivers during the earlier
part of the period we are considering would also tend to
the more frequent occurrence of floods, the flats on which
coal was growing would be more liable to be inundated by
muddy water, and the coals in consequence would be more
earthy and impure than those formed during the period
of the Middle Coal Measures.

There now remains for consideration only the topmost

subdivision of the carboniferous rocks, the Upper Coal
Measures.

The rocks of which the group is made up are mainly
shales and sandstones; it contains seams of coal, but they
are neither so numerous nor so thick as those of the
Middle Coal Measures. Thin bands of limestone are also
met with in this subdivision.

Their poverty in coal and the occasional occurrence in
them of a few insignificant layers of limestone furnish
but slender grounds for separating these beds from the
Middle Coal Measures, and it is mainly on the strength of
their peculiar colour that they are erected into a distinct
subdivision.

The bulk of the rocks of the Middle and Lower Coal
Measures and of the Millstone Grit are blue or grey where
protected from the action of the air, and of various shades
of brown or yellow at or near to the surface; the sandstones
of these subdivisions for instance are nearly always of the
latter colour in quarries, unless the quarry be very deep,
when the lower beds assume a greyish hue; at moderate
depths the blocks of stone are not unfrequently grey inside, 'grey-hearted' as the phrase is, with an outside brown
or yellow crust. When brought from greater depths, as
in colliery sinkings, the rocks are always grey or blue.

The grey or blue colour is probably mainly due to organic
matter. The change in colour is owing to the fact that
in the grey or blue rocks there is iron in the state of
ferrous carbonate $(FeCO_3)$; this salt when exposed to the
oxidising action of the atmosphere is converted into a
brown or yellow ferric hydrate $(Fe_2O_3 + water)$, the
colouring power of which is intense enough to mask the
original greyish hue of the rocks. The exact tint which
results is probably dependent on the degree of hydration.

The prevailing colours, on the other hand, of the

Upper Coal Measures are red, purple, and mottled red and green.

This difference in colour may seem at first sight to be but a feeble reason for placing the red beds in a separate subdivision, but a change in colour, no less than a change in mineral character, implies a corresponding change in the conditions under which the beds were deposited, and we may safely infer that a group of rocks which are markedly red was not formed under the same circum- stances as a body of rocks uniformly blue. We will now enquire what are the conditions suitable for the formation of red rocks like those of the Upper Coal Measures, and we shall find that they differ in one respect so widely from the conditions prevailing during the deposition of the Middle Coal Measures as fully to justify the separation of the two groups.

The red colour of the rocks we are now dealing with is caused by every grain being coated by a thin skin of ferric oxide (Fe_2O_3). This may easily be shown to be the case by crushing a bit of one of the red sandstones and boiling it in hydrochloric acid; the colouring matter is dissolved, and there remains a quantity of sand of the purest white colour. Every grain then of the sediment out of which these rocks were formed must have obtained, before it came to rest on the bottom of the water, a thin coating of this red colouring matter, and in no other way could this result be brought about but by the colouring matter being distributed in large quantity throughout the whole body of the water itself. The water in which these red rocks were formed was itself red as blood owing to the presence in large quantity of very finely divided ferric oxide.

Now clearly such a state of things could never come about in a lake or estuary like that in which the Middle

Coal Measures were deposited. However much red colouring matter was brought in at one end, the bulk of it would
be carried out at the other end, for the feeblest current
would suffice to sweep on matter in so fine a state of
division. But close up the outlet of the lake, and the
conditions necessary for the production of blood-red water
follow as a matter of course. Water plus colouring matter
is brought in, water is removed by evaporation, but no
colouring matter is thereby abstracted, the amount of
colouring matter therefore increases day by day, till the
whole body of the water becomes thoroughly saturated
with it.

And when we turn back to the history of the Middle
and Lower Coal Measure period, we find that the closing
of the outlet is a thing that might not unreasonably be
expected to occur. The marine bands, each of which
marks an incursion of the ocean outside, seem to become
less numerous as we ascend in the measures, so that in all
probability the barrier that held back the sea grew firmer
and firmer as time went on, and its gradual increase in
strength may in the end have resulted in the strait which
connected lake and ocean being completely blocked up.

The limestones tell the same tale. They are not, like
the limestones of the Carboniferous Limestone, made up
of the hard parts of animals; they have a peculiar creamy
look when freshly broken, they are sometimes finely
banded, and occasionally they have a porous spongy
texture.

Limestones not produced by organic agency can be
formed only in one way, by precipitation from water
holding carbonate of lime in solution, and the deposits
formed in this manner possess very generally the peculiarities which characterise the limestones of the Upper Coal
Measures. In the incrustations which coat the walls and

spread over the floors of caverns and in the long pendants which descend from their roofs, known as stalagmites and stalactites, the slow deposition of film upon film of mineral gives rise to a banded structure; and in many of the deposits formed by calcareous springs, those of Matlock and Tivoli for instance, the escape of gas has filled the mass with cavities and winding channels that give it a spongy texture.

The limestones of the Upper Coal Measures may then be fairly looked upon as precipitates from water holding carbonate of lime in solution, and they indicate just in the same way as the abundance of the red colouring matter, that the body of water in which they were deposited had no outlet. For precipitation will not take place till the water is saturated, that is, till it holds in solution as much as it is able to hold. Now saturation is, to say the least, very unlikely in a lake with an outlet; but, given time enough, it must be arrived at in a closed body of water into which mineral springs discharge themselves, because, while evaporation is always drawing off water, it carries away none of the matter in solution.

It may now very fairly be asked where the red colouring matter and the water impregnated with carbonate of lime came from. It is probable they came from a volcanic source. Ferric oxide, under the form known as specular iron ore, is frequently found coating the fissures through which fumes discharge themselves in volcanic districts. This substance has probably been produced by the decomposition of chloride of iron, which rises in a state of vapour from the heated depths. Again volcanic districts abound in mineral springs, and of all mineral springs calcareous springs are the commonest.

And volcanos were not far to seek. What are now

F

the Lowlands of Scotland were at this time thickly dotted over with volcanic cones, the sites of which can be even now fixed. Elsewhere in the British Isles volcanos of Carboniferous date have left their marks behind them. The Carboniferous Limestone of Derbyshire, the Isle of Man, and Ireland includes subaerial flows of lava and beds of volcanic ash, and the Whin Sill of the North of England and the ' Green Rock ' of the South Staffordshire coal-field are sheets of lava that burrowed underground among recently formed deposits of Carboniferous age.

With volcanos close at hand, and with the known habit of volcanos to yield incrustations of oxide of iron and to give rise to mineral springs, we may reasonably refer the red colour and limestone bands of the Upper Coal Measures to a volcanic source.

It may be further asked why were red colouring matter and carbonate of lime furnished in such special abundance towards the close of the Carboniferous period. In reply it may be said that it is not certain that the yield was then more plentiful than it had been heretofore, but that, after the lake had become closed, all that came was caught and retained. But it is not unlikely that during the deposition of the Upper Coal Measures the supply of the substances named was somewhat exceptional. The volcanos just mentioned were the last expiring efforts of a long course of volcanic action, and it is when volcanic energy is on the decline or approaching extinction that Solfataras are most active and mineral springs most abundant; indeed, long after all the more violent demonstrations have ceased, these remain as witnesses of the former presence of volcanic activity of a more energetic kind. It was in fact just during that phase in the life of these old volcanos when they were most likely to furnish matter that would colour rocks red and most likely to

give rise to calcareous springs, that the red rocks and
the limestones of the Upper Coal Measures were deposited.

One more point in the history of the Carboniferous
rocks has yet to be touched upon and we have done.

The map on fig. 3 shows that originally an unbroken sheet of coal-bearing rocks must have spread over
nearly the whole of the North of England, the Lowlands
of Scotland, and Ireland : but of this vast sheet only fragments now remain.   Turn to a geological map of the
British Isles, and among the various colours with which
that map is diversified sundry patches of a deep black
tint will be noticed.   These mark the areas in which
coal-bearing rocks now occur at the surface, the coalfields
as they are usually called.   A large part of the space that
separates these black patches is occupied by Millstone
Grit or Carboniferous Limestone.   Coal Measures were
once present above these rocks; but they are there no
longer..   The extent to which the Coal Measures have
been removed is very large in England and Scotland, but
Ireland has suffered more severely still, and she has
scarcely anything left of the broad sheet of coal-bearing
beds 'that originally covered nearly the whole of her
surface.

How has all this havoc been wrought?   The first
fact that helps us to an answer is this.   All rocks
formed under water must have been deposited in layers
that were originally horizontal or all but horizontal; but
this is not the position in which we now find the beds of
the rocks we are considering; they slope or dip at various
angles to the horizon.   They have therefore been tilted
since the date of their formation out of the horizontal
position in which they were laid down.   Further, when
we come to examine the lie of the beds over large areas,

we find that this tilting was caused by the rocks being folded into a succession of broad troughs and arches.

Now think what would happen if a group of beds lying flat beneath water were subjected to this treatment. The crest of each arch would gradually rise to the surface of the water; when it was near that level the waves and currents would eat and wear its summit away, and as the upward movement went on, slice after slice would be planed off. Suppose at last the truncated top was lifted beyond the reach of the waves, and converted into a land surface. Rain and running water, frost and other agents then attack it, and carry on the work of destruction. In this way the higher portions of the rock group are swept off the summit of each arch, and the lower members laid bare.

To denude is a Latin word which means to lay bare, and the process by which the lower beds are stripped of the covering, beneath which they were originally hidden, is called ' denudation.'

The troughs or depressions between the arches are less exposed to the wear and tear of denudation than the summits of the arches, and in them the upper beds still survive.

Each of the black patches of Coal Measures is more or less of a trough or basin; while in the intervening spaces, where the Millstone Grit and lower rocks come to the surface, the beds are folded over in arches, in the way shown in the section in fig. 4.

At first blush it may seem as if Nature had treated us rather hardly in leaving us so small a remnant of the vast store of mineral wealth which the country once possessed. But it is not so. But for this tilting and denudation every ton of coal would have been buried at depths so enormous that even if we had known of its existence, it would have been utterly beyond our power to reach it. The thick seams of the Middle Coal Measures would have had above them not only the whole thickness of the Upper Coal Measures, but many a thousand feet of rocks of later date as well. No ; here, as everywhere if we can only see into her reasons, Nature has proved herself a kind and provident mother, taking away a part in order to place the remainder within our grasp.

Such then was the process by which the great sheet of Coal Measures was broken up, and the surviving portions were thrown into the form of basins.

It must not be supposed however that the black patches on the map represent all the Coal Measures that are left to us.

These show the parts of each basin where the Coal Measures actually come to the surface, which are generally distinguished as the ' exposed coalfields.' In some cases rocks of later date overlie a part of the basin, and the portions hidden beneath these rocks are often spoken of as ' concealed coalfields.' In fig. 4, the whole of the basin on the left is ' exposed,' but of the basin on the right the right hand half is ' concealed.'

And now let us gather up, in as few words as may be, the story that has been told at length in the preceding pages.

The north of England and the south of Scotland were occupied at the opening of the Carboniferous period by a land-locked sea which was connected on the west with the ocean by one or more outlets of no great breadth. In this sea the Carboniferous Limestone was deposited. The accumulation of the Carboniferous Limestone gradually filled up the deeper parts of the sea and gave rise to a tract of shallow water in which the muddy Yoredale Rocks were formed.

FIG. 4.

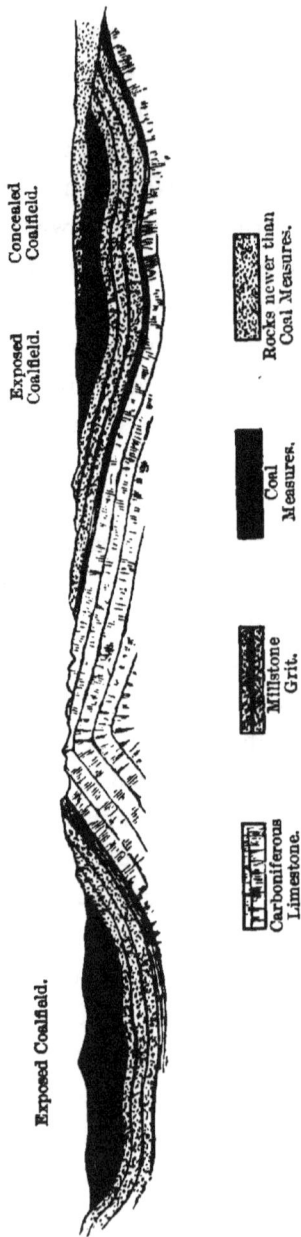

Exposed Coalfield.

Exposed Coalfield.

Concealed Coalfield.

Carboniferous Limestone.

Millstone Grit.

Coal Measures.

Rocks newer than Coal Measures.

A barrier was then established across the outlet which held back the outside ocean, and the inland sea was transformed into a fresh-water lake or estuary in which the Millstone Grit and Lower and Middle Coal Measures were formed. The bulk of these subdivisions consists of rocks deposited in water of no great depth, and the large aggregate thickness which they reach was rendered possible by the bed of the water sinking slowly while they were accumulating. Every now and then the sinking stopped, the water became filled up by sediment, and swampy flats were established on which materials for coalseams accumulated by the growth of land-plants. The barrier that held back the sea was occasionally breached, and salt water gained access to the area. During these incursions bands containing marine fossils were deposited.

At last the outlet became completely blocked up, a closed body of water was formed, and in it the red Upper Coal Measures were deposited.

Finally the great sheet of rock thus formed was crumpled up into huge arches and troughs, the higher members were denuded off the crests of the arches, and the remnants of the Coal Measures were preserved in basins.

My tale is finished, and in conclusion I would ask whether I have made it appear that the methods which geology uses to decipher the story written on the rocks of the earth's crust, are sound and good.

If I have been successful in vindicating geological reasoning and geological conclusions, I think I may fairly say that the geological record, fragmentary and hard to read as it is, yields neither in interest nor accuracy to any of the histories of more recent times.

It includes none of those painful interludes which mar

so constantly our pleasure during the perusal of the history of our own race. It is concerned with the majestic revolutions of nature, and are not they pleasanter to contemplate than those instances of man's littleness, misery, and cruelty that stain so thickly the page of human history? And when we take into account how all but impossible it is to eliminate the countless sources of error that beset the interpreter of the chronicles of the human epoch, shall we dare to say that records penned by fallible and biassed mortals are more trustworthy than those traced by the unerring hand of Nature herself?

CHAPTER III.

COAL PLANTS.

MORE than one writer has commented upon the difficulty
of arriving at any certain conclusion respecting the affinities of a fossil plant. The object of inquiry is generally
most fragmentary, and the truly significant details are
seldom seen. A leaf-pattern is perhaps disclosed in pleas-
ing but useless perfection; a stem is made to reveal some
part of its internal structure by the patient toil of the
lapidary, but the form and arrangement of those organs—
generally deep-seated, nearly always perishable--which
determine affinity according to the systems of modern
botanists, are in most cases hardly to be guessed at. The
great botanist Robert Brown, as Sir Roderick Murchison
relates, 'never would pronounce upon the genus, scarcely
even upon the class, of a fossil plant,'[1] while Sir Joseph
Hooker, himself an eminent investigator of Coal Measure
plants, has written many a page to show with what
caution we should receive the identifications of vegetable
palæontology.

This scepticism is not, it may well be supposed,
groundless, but the reproach which gave it a sting is less
deserved now than formerly. When men first turned to
the study of fossil plants, they were in haste to name and
classify; systems sprang up as in a day; identifications
and discriminations cost hardly an effort; nor were
inferences respecting the climate of ancient epochs, or the

[1] Geikie's *Life of Murchison*, i. 214.

course of life upon the earth, less plentiful. But the tone
of recent workers is very different—much cooler, less
attractive, more promising of safe results. Relinquishing
the attempt to give to every fossil a name and definition,
the palæontologist now aims rather at working out in
detail the structure of a few types or of one. This method,
slow and tiresome as it is, has been productive beyond
expectation. Exploring the immeasurably ancient flora of
the Coal Measures, the oldest of any completeness known
to us in Europe, palæontology has in our own time
brought to light the most delicate organs and tissues, and
has founded thereupon relations of affinity which can bear
the criticism of botanists trained by the exact comparison
of fresh and perfect plants. It could not now be said that
the study of the plants of the Coal Measures is inferior,
either in respect of accuracy or importance, to those
investigations of fossil animals with which it has been
disadvantageously compared.

I shall endeavour in the following remarks to exem-
plify the method, as well as to quote the results, of recent
work. A somewhat partial treatment thus becomes
necessary; the space, which equally subdivided would
suffice only for a meagre sketch, may be large enough for
the discussion of all those details concerning one or two
types, into which an unprofessional student can enter.
The choice of types is not difficult. There are two groups
of Coal Measure plants, so plentiful, so peculiar, and so
characteristic of the formation, as to demand the chief
consideration. These are the Calamites and the Lepidodendroid trees, of which latter group Lepidodendron and
Sigillaria are the principal examples; and for these I
solicit the greater share of the reader's attention.[1]

[1] The following sketch of the two leading types of Carboniferous
vegetation is taken either from the common stock of botanical and

One of the commonest fossils of the Coal Measure sandstones is a jointed and furrowed cylindrical cast, which when perfect tapers to an obtuse point at one end. The diameter of the uncompressed cast seldom exceeds four or five inches; the joints (nodes) are often three or four inches apart ; while from nine to ninety longitudinal furrows may be counted round the surface. Towards the pointed end the nodes are often closely crowded together; the apex may be straight, but is more usually curved. At the upper end of each internode (space between two nodes) there is often seen a circle of roundish scars, one to each longitudinal ridge. Occasionally a large depressed scar, indicating the attachment of a lateral branch, is seen at the junction of two internodes. When the plant occurs under such conditions as favour the preservation of structure—when, for example, it is imbedded in a fine homogeneous shale or ironstone instead of sandstone—it may happen that an outer coating of coal, extremely like the ordinary Calamite in its superficial marking, invests the cast. In this case the longitudinal furrows as well as the nodes of the two surfaces generally correspond very closely, both as to number and position. Such are the more obvious external characters of the fossils known as Calamites. They were compared by the earlier students of fossil plants to reeds or bamboos—

palæontological knowledge, or from the invaluable memoirs of Prof. W. C. Williamson, F.R.S. (*Phil. Trans.* 1871, et seq.). Few will question the statement that this splendid collection of researches is entitled to the highest honours in a field where many have attained to excellence. Prof. Williamson has allowed me to study the preparations of Carboniferous fossils upon which these memoirs were founded, and has removed many difficulties by his oral explanations. It will therefore be understood that the chief responsibility for the facts of structure here related rests with Prof. Williamson, while I am to be held answerable for the way in which the abridgment is executed.

hence the name; most authors now assign them to the horse-tails or Equisetaceæ.

When Calamites were first described and figured, the furrows and constricted nodes were supposed to characterise the outer surface of the plant, the tapering end was very naturally regarded as the growing point, and the circle of scars adjacent to every node was taken to indicate the attachment of a whorl of leaves to the stem. All these suppositions prove to be incorrect. It is only in our own day, and by the patient labour of many workers, that the most fundamental questions of structure have been satisfactorily answered. Well preserved stems, imbedded, not in sandstone, but in ironstone or calcareous nodules, have been sliced by the lapidary's wheel, and from the careful study of hundreds of these sections, a general conception of the organisation of the plant, accurate and useful, though far from complete, has been elaborated.

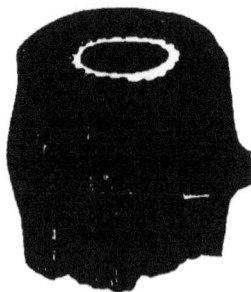

FIG. 5. Restoration of part of stem of young Calamite, showing a node and portions of two internodes. From Williamson ('Phil. Trans.' vol. clxi., pt. 2, 1871).

Let us first examine the structure of a single internode. Each contains a number of vertical radiating wedges, separated by corresponding plates of pith-like cellular tissue. The wedges consist of bundles of long vessels intermixed with cellular tissue after a fashion

which will be immediately described. The entire mass
of wedges forms what is known as the 'woody zone,' but
it should be remarked that the common form of wood-
cells, met with in our forest-trees, has not been found in
any Calamite.

If it be now supposed that the hollow cylinder just
described as the 'woody zone' is invested by a bark and
lined by a thin layer of central pith, while the vertical
radiating plates of cellular tissue (the 'primary medullary
rays' of Prof. Williamson), which separate the woody
wedges, are continuous with both pith and bark, the leading features of the internal structure of a young Calamite
stem will be apprehended. We now pass on to examine
the details of the arrangement.

A solid pith is rarely met with, and only in very
young stems. As the stem enlarged, the walls of the
central cavity became more remote, until the pith, unable
to stretch or to resist tension, broke asunder in the
centre, leaving a gradually increasing space within each
internode. In many of the younger but moderately
thick stems, the pith is found lining the woody cylinder,
sending vertical radiating plates (the primary medullary
rays) between the woody wedges, and forming a more
or less complete transverse diaphragm at each node. In
old stems the pith may be completely absorbed. Prof.
Williamson observes that the occasional perfect preservation of individual pith-cells in hollow stems and
the almost geometric regularity of sandstone casts
moulded upon the sharply defined and very thin lining of
pith show that the central cavity is due to disruption and
not to decay. This conclusion is fortified by a different
set of facts which will be related further on.

The primary medullary rays extend vertically from
node to node, and radially from pith to bark. They

consist of elongated cells without internal thickening, regularly disposed in rows like the bricks of a wall, and hence termed 'muriform tissue.' To make the comparison perfect, the bricks should be set upon their narrow ends, a single row in thickness, but in several vertically superposed courses.

The bark of a Calamite is very rarely seen. In the few young stems where it has been detected, it forms a thick and almost uniform envelope, which presents when cut across no separation into histologically differentiated layers. No furrows or surface-ornaments are known to characterise the exterior of the bark, nor are the transverse constrictions corresponding to the nodes seen upon it, as upon the casts of the pith-cavity. In an old stem the bark, which was very thick, was probably cracked in all directions by the enlargement of the internal structures. It became differentiated into a thin inner layer of ordinary cellular tissue (parenchyma), and a thick outer layer composed of long, pointed cells (prosenchyma).

The general disposition of the woody cylinder has already been noted ; we proceed to analyse a single woody wedge. In its earliest state each wedge is indicated by a single apical canal, passing vertically from one node to another along what is subsequently to form the sharp internal angle of the wedge. To the outer side of this canal are gradually added laminæ, radiating outwards and increasing in number by intercalation. Each lamina consists of a row of vessels, standing upright like a rank of organ-pipes, the different laminæ being kept asunder by vertical cellular partitions (the 'secondary medullary rays' of Prof. Williamson), which are held to correspond to the medullary rays of the higher exogens. A sufficient magnifying power reveals further details. Each vessel of a row is squarish in section, differing thus from the round apical canal, which is also of larger size. The vessels exhibit not a little resemblance to the ' scalariform ' vessels of a fern. Each is strengthened on its internal surface by cross-bars, due to a deposit of hard tissue upon the delicate cell-wall, but these bars do not, as in true scalariform vessels, pass into vertical ribs at the angles. The secondary medullary rays consist of muriform tissue, the cells having the same disposition as those of the primary medullary rays. When a tangential section of a wedge is made (as when a plank is sawn off the outer surface of a tree) the secondary medullary rays are seen compressed between the laminæ, and their narrow, thin-walled cells are readily distinguished from the thick ' barred ' vessels.

A young Calamite exhibits in the arrangement of its tissues a noteworthy resemblance to a young exogenous stem of the first year. In both we find a circle of detached wedges interposed between pith and bark ; in both the formative area is at the outer surface of the woody cylinder; in both a system of primary and secondary medullary rays is traceable. In living exogens the fresh-formed product of the cambium-layer, immediately internal to the bark, consists of two sets of cells—those formed in spring, and those formed in autumn. The autumnal zone is denser, and contains fewer vessels, and to this difference the rings seen in a cross-section are due. The Calamite stem (and the same holds equally good of Lepidodendron) is exogenous, but not ringed. This may be an effect of climate ; concentric rings are rare even in the coniferous Dadoxylon of the Coal Measures.

The description of pith, bark, and wood already given sufficiently explains the structure of a single internode of the Calamite. The entire stem is made up of many superposed internodes. It is necessary to consider them in their relation to each other, and in particular to describe the rearrangement of fibres which takes place wherever two internodes meet.

At each node in very young stems, the woody cylinder undergoes a temporary increase of diameter. This is due to additional longitudinal vessels which arch over the node. The innermost vessels are very short, the outer ones long enough to meet the corresponding set from the nearest nodes above and below. This slight nodal dilatation was probably apparent in such young stems, not only upon the woody cylinder, but (less distinctly) upon the outer or cortical surface of the stem. The apical canals, traversing the internal angle of each woody wedge, sometimes end blindly at the nodes, though they may in other cases subdivide and be prolonged into the adjoining internodes. At the same points the cellular sheets, known as the primary medullary rays, become nearly obliterated, remaining only as somewhat enlarged ordinary rays.

The barred vessels of each woody wedge subdivide at each node into two wedges, thus forming a series of arcades radiating from the pith, each lateral half of a wedge uniting with the adjacent half of the next wedge of the same internode so as to form what an anatomist would call a chiasma. Such united bundles form the woody wedges of the next internode.[1] This rearrangement of the woody cylinder explains the almost complete obliteration at the nodes of the primary medullary rays, and the regular alternation in successive internodes of the woody wedges. Another important structural detail is sometimes revealed by tangential sections at or above the nodes. The woody bundles may be pierced, just at the points from which the nodal arches spring, by oval spaces, and

[1] Compare fig. 6, D.

through some of these spaces pass other and horizontal bundles of vessels, derived from the inner layers of the woody zone and destined to supply a branch. The vessels in one of these perforating bundles increase in number as the tangential section approaches the periphery of the woody cylinder. They take at first an uncertain, meandering course, but gradually arrange themselves in the direction of the branch, and when cut across, at some distance from their origin, exhibit the same radiate disposition as the vessels of the main stem. A lucky radial section may show a prolongation of the pith traversing the centre of the outward-directed bundle. In very young stems the slender branches were probably whorled, since one of these oval spaces sometimes perforates each woody wedge, and an approximation to this number is very often met with; but many branches abort or fall off, so that in an older stem the spaces transmitting bundles are fewer than the wedges. In a still older stem there is often only a single branch to several nodes, and this of so large a size that the transmitting space loses its original character altogether, and an extensive rearrangement of the surrounding tissues occurs. The spaces in question differ altogether in character from the 'infra-nodal canals' shortly to be described, and must be carefully distinguished from them. They differ also in position, for these infra-nodal canals, as the name implies, are always found immediately below a node, whilst the spaces corresponding with the insertions of branches lie in or above the node.

In an old stem the primary medullary rays gradually disappear, and the consequent closing up of the vascular wedges effaces the rearrangement of their lamellæ seen at the nodes in younger examples.

It should be noted that the primary woody wedges are

not certainly known to increase in number when once formed. The apical canals, which precede the formation of the wedges and indicate their future position, appear to be very nearly always disposed at equal distances and in a regular circle. This seems to show that this ring of canals is formed, not successively, but simultaneously. It is however observed that young stems have frequently a smaller number of wedges than older ones—a fact difficult to explain. The laminæ of each wedge continually multiply by intercalation. At first there is the apical canal without laminæ; then a few short laminæ are seen radiating outwards from the canal; while in an old stem very many laminæ can be distinguished upon the outer or broad end of the wedge.

Prof. Williamson has described under the name of 'infra-nodal canals' certain passages which are frequently indicated upon internal casts by a circle of scars, one to each longitudinal ridge. These infra-nodal canals radiate towards the bark like the spokes of a wheel, and appear to have formed channels of communication through the woody zone between the top of each internodal cavity and the inner surface of the bark. It is not known that they penetrated the bark. Their physiological significance is very uncertain. It appears that they arise by the constriction and ultimate detachment of the upper portion of a primary medullary ray, so that they are merely special modifications of parts of these cellular tracts. Recent observations have shown that they in some cases retain their cellular contents unabsorbed, and thus become radiating spokes of a specialised cellular tissue connecting the pith with the bark.

Turning next to the habit of these plants, we have reason to suppose that many aerial stems ascended from a branched and creeping subterranean rhizome, much as

in the recent horse-tails. The basal part of each aerial
stem gave off roots, always from the lower part of an internode, and these have been identified with some of the
long slender shoots, densely clothed with radiating capil-
lary branches, named Pinnularia. Lateral aerial stems
were frequently given off from the base of a main trunk;
their position is often indicated in the medullary casts by
large roundish sunken scars at the nodes of the parent
stem. The insertion of a lateral branch, as occasionally
seen in a Calamite preserved in shale or sandstone, is at
first sight not a little puzzling. The branch seems just
to touch by its pointed end the centre of the rounded
scar, all its weight apparently depending upon a thread.
It is plain at a glance that this could not have been all.
In reality the fossil is only a cast of the pith-cavity. All
the tough and solid surrounding tissues, which formed
the true supporting tissues, have rotted away, leaving
only the mark of their attachment upon the parent-stem.
It has already been pointed out (p. 81) that the whorled
branches, present in an early state of growth, rarely per-
sist, and that an old stem gives off only a branch here and
there, not even one from every node, but at much greater
intervals. The permanently surviving branches attained
a good size, even as compared with the main trunk, from
which in the state of medullary casts they are often only to
be distinguished by the curved (instead of straight) conical apex. A substantial junction existed between the
bark and woody cylinders of trunk and branch, while the
pith-cavities communicated by a continuous (originally
cellular) passage, leading into the upper part of an internode of the parent stem.

We have already remarked that the Calamite usually
occurs in the form of an inorganic cast, and it may
be convenient that we should now apply the ascer-

tained facts of structure to the systematic interpretation
of the fossil in its common state of preservation in shale
or sandstone. The cast is usually a filling-in of continuous
internodal cavities. It has been pointed out that examples with a straight conical base probably represent the
primary aerial stems springing from a subterranean
rhizome, while the lateral aerial stems have curved conical
bases, and numerous crowded nodes. Scars representing
the infra-nodal canals are often seen in a circle immediately
below a node, and the upper and lower ends of a fragmentary cast may be distinguished thereby. The flutings of
one internode alternate with those of the next. It will
be remembered that the fluting is the expression of the
primary medullary rays, and that these are shifted at
every node, each half its own breadth, in consequence of
the redistribution of the woody bundles. The superficial
markings of a sandstone cast are very much the same,
whether it is the mere filling-in of an internal pith-cavity,
or whether the carbonaceous film, which represents the
vegetable cylinder, is present upon its surface. This has
been interpreted to mean original identity in surface-
marking of two concentric cylinders, and has given rise
to perplexity and mistake. The close resemblance is apparently due to the moulding of the coaly film upon the
previously solidified cast of the pith-cavity. Did the
ridges and furrows of the coaly film represent in reality the
exterior markings of the woody cylinder, then, instead of
corresponding with the ridges and furrows of the medullary cast, they would have alternated therewith. The
external ridges would be bulgings outwards of the woody
wedges themselves—the internal ridges interspaces between the wedges ; for it must be remembered that where
each woody wedge projects most in the direction of the

pith, it is also equally prominent at its cortical surface, i.e. at the centre of every wedge.

It has been contended by naturalists of eminence, such as Brongniart, Dawson, and Binney, that the Coal Measure Calamites are divisible into two perfectly distinct, but outwardly similar types. One of these (*Calamodendron*) has been held to be a gymnospermous exogen, allied to the fir-trees, while the true *Calamites* is regarded as Equisetaceous. Prof. Williamson believes that the distinction depends entirely upon the state of growth and preservation, and that a Calamodendron is merely a Calamite which retains its internal structure, and especially its exogenous vascular zone. The organisation of the stems attributed to Calamodendron explains all the details of the fossil casts, while no fragment revealing minute structure, from whatever locality derived, can be assigned to the restricted genus Calamites.

It has been conjectured that the branches named Asterophyllites, Annularia and Sphenophyllum, with their whorled leaves, represent the leaf-bearing branches of Calamites. The foliage of the Calamite was in all probability similarly verticillate, and there is reason to expect that it may hereafter be identified with some of the forms just named, but proof positive, such as would be supplied by the discovery of the leaves actually attached to an unmistakable Calamite stem, is not so far accessible.

Our information respecting the organs of fructification is also, but in a less degree, imperfect. Analogy of allied plants, recent and fossil, would lead us to suspect that the Calamite fruit consisted of a spike of modified leaves, lodging sporangia or spore-cases. Such spikes, unappropriated to other plants, do occur in the Coal Measures, and have been carefully studied. But of many different types known to exist there seems to be one only which can be associated with the present genus on the clear evidence of attachment to an indubitable Calamite stem. The one British example of this type[1] shows a central stem occupied by a comparatively large medullary cavity, and sending out from each node a whorl of bracts. Every bract gives off two upward directed spines (sporangiophores or spore-case bearers), each of which is surrounded by a compact mass of spore-cases. Within the spore-case are many single-celled spores. The central stem exhibits an internal structure which leads Prof. Williamson to identify it with the Calamites.

The facts of structure now laid before the reader lead to the conclusion that the Calamites were Cryptogamic plants, that is, that they were reproduced, not by flower and seed, but by spores. We shall have occasion to examine more closely the precise nature of this distinction when the fruit-cones of Lepidodendron come under consideration. In its relation to living Cryptogams the Calamite is exceptional in the complex structure and exogenous growth of its woody cylinder, increasing as it does by indefinite additions to the exterior of each of the radiating wedges; but it will be seen hereafter that this structure and mode of growth are repeated in other Carboniferous Cryptogams, while they are imperfectly paralleled in one recent plant, Isoetes, which belongs to the same great division.

The Equisetaceæ, or horse-tails of our fields and marshes, bear marks of resemblance to Calamites, far inferior though they are in stature. They have the creeping, subterranean rhizome, and the furrowed, fistular stem, with diaphragms of pith, in which an arrangement of tissues comparable to that of the Calamite, though less

---

[1] See Williamson on a new form of Calamitean Strobilus, *Mem. Lit. and Phil. Soc. of Manchester*, 3rd ser. vol. iv. pp. 248–265, pl. vii.–ix. (1871).

complete and less persistent, may be traced.   The fruit-
cone is not unlike that assigned to the Calamite, though
it is more special, more profoundly modified from the

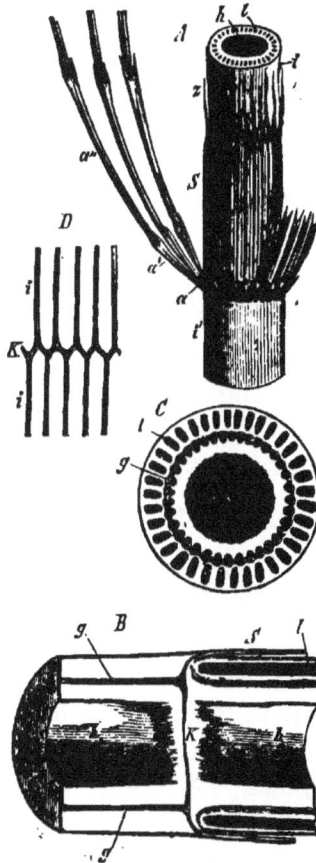

FIG. 6. Recent Equisetum. A. Piece of upright stem. i, i', internodes;
h, central cavity; s, leaf-sheath. B. Longitudinal section of underground stem (magnified). k, septum between the cavities of the
internodes h, h; s, leaf-sheath. C. Transverse section of the same.
D. Union of fibro-vascular bundles at the node K. From Sachs'
'Botany.'

primitive leaf-branch.   Recent Equisetaceæ differ in two
notable respects from Calamites.   They have a leaf-sheath
investing every internode, which springs from one node
and splits into a whorl of leaves at the node next above.

Again the spores of the recent horse-tails are each provided with four elastic threads (elaters), which spread out
in a cross when dry, but curl spirally round the spore
when damped, and often so suddenly as to set the tiny
globule rolling, When spores of an Equisetum strewn
over the field of the microscope are lightly breathed upon,
they are excited to active motion by means of their
hygroscopic threads. No such appendages are known to
belong to the spores of any Calamite, or indeed to any
fossil spores.

The classificatory value of these resemblances and
differences is variously estimated by authors. Whether
Calamites should be regarded as a peculiar form of
Equisetaceæ, or as belonging to a distinct but allied order,
is a question which we shall not attempt to decide. For
the purpose of summary statement we shall provisionally
treat them as an outlying type of Equisetaceæ, observing at
the same time that they cannot be actually incorporated
with the recent horse-tails without a fresh definition of
the order.

A group of Coal Measure plants, in many respects
more important than the Calamites, consists of those
known as Lepidodendroid or Lycopodiaceous. To this
group are referred Lepidodendron, Sigillaria, and others
less familiar. They are believed to be similar in their
main features to the club-mosses (Lycopodiaceæ) now
living in this and many other countries. The fossil club-
mosses are not however identical in structure or outward character with any recent species; in particular,
they are much larger, often rising into trees, while the
club-mosses of our own time are all of small size.

Hitherto the leaves and pieces of bark imbedded in
coal shale or sandstone have been arranged and named
almost exclusively by outward pattern. It is necessary to

warn the beginner not to trouble himself with the crowd
of specific names which have thus arisen. Different parts
of the same tree have been named as separate species,
according as the outer surface of the bark, or an impression of one of the inner layers was exposed to view. A
parallel case quoted by Dr. Balfour [1] will show how easily
this mistake may be made. An Araucaria 24½ feet high
was destroyed by frost in the Edinburgh Botanic Garden.
The external surface of the bark was covered by polygonal
scars spirally disposed, each with a narrow ellipse in the
centre indicating the insertion of a leaf, and surrounded
by radiating prolongations. When the epidermis or
outer bark was removed, a notable difference appeared
between the polygonal scars towards the bottom of the
trunk and the lozenge-shaped scars higher up. Both
patterns differed from the superficial marking of the
epidermis. The middle bark also presented a great variety
of forms, but this may possibly be attributed to the
effects of frost upon the cells. Mr. Binney has found a
single Sigillarian stem to combine the patterns attributed to four species.

It would hardly be possible to describe in a single
sentence the many patterns which a Lepidodendroid bark
may assume. In Sigillaria some kind of vertical fluting
is common, and the leaf-scars are often arranged in pairs
at regular intervals and in vertical rows. In Lepidodendron the leaves densely clothed the young twigs, but
shrivelled up and left the older stems bare. After they
fell off, scars still indicated their former insertion. The
scars enlarged as the stem increased in diameter, just in
the same way that a name cut in the bark of a young tree
enlarges from year to year. The Lepidodendron stem is
completely covered in this way by old leaf-scars which

---

[1] *Palæontological Botany*, p. 4, figs. 2–5.

have expanded till they meet. The pattern is quincuncial $(*_*^*_*)$ and each scar is lozenge-shaped. It is probable that in very old stems the scars entirely disappeared.

Sigillaria is as yet very imperfectly known. It seems to be established that the roots were similar to those of Lepidodendron, and that the internal structure of the stem was in its main features the same. The upper part of the tree is not known for certain, for singularly enough neither branches nor young stems of Sigillaria have been found, and it cannot be said whether Sigillaria had the repeatedly forked branches characteristic of Lepidodendron. Nor is much directly known of the foliage or fruit, but there is a strong antecedent probability that they were generally like those of Lepidodendron.

Seeing how imperfectly we are acquainted with the structure of Sigillaria, it is obviously preferable to take Lepidodendron as our type of the Carboniferous Lycopodiaceæ. Remains of the plants which bear this name are among the commonest coal-plants, while the belief that they have in some cases furnished an important constituent of Coal itself, renders an examination of their organisation the more necessary.

Taking first into consideration the stem and the roots of the Lepidodendron, we find several features common to nearly all trees, whatever their botanical affinities. Similar physiological needs impose similar adaptations upon plants of very diverse ancestry. The function and purpose of a tree is to rise above the low jungle, crowded with herbaceous and trailing plants, into the sun and air. Height gives advantage, and far beyond the limits of the advantage originally sought, competition requires the forest trees to go on rising and rising, that they may overtop, and not be overtopped. A copious mass of foliage, exposing thousands of square feet to the air,

must be supported and extended. Then there are reproductive elements—seeds, spores, pollen—to be scattered to
the wind. Lastly, the whole fabric, lofty and spreading,
must be held fast in the earth by strong and well-set
roots. These are with hardly an exception the conditions
of life for all trees. We find accordingly a certain wide-
spread, though not universal, similarity of structure. In
the stem of Lepidodendron, as in a fir or an elm, there is
a softish bark, growing rapidly when young, extensible,
conducting heat badly, checking evaporation; there is a
cylinder of wood capable of increase by external additions;
there are vessels for the transmission perhaps of liquids,
perhaps of air; there is a central pith. But the original
diversity of the plants, which have so many physiological
needs and physiological adaptations in common, comes
out when we study the details of their structure.

The roots of both Lepidodendron and Sigillaria are
extremely similar, and received the common name of
Stigmaria before their real nature was ascertained, and
when they were supposed to constitute a distinct plant.
The main stem often divided in or upon the ground into
four principal roots (fig. 7), which extended themselves
horizontally like the arms of a cross.[1] Examples of the
entire root-mass are occasionally met with, and these
have been known to show on the under-side a kind of
seam or suture, which marks the insertion of each main
root upon the stem. The four lines of suture may simply
decussate, or the two lines of one side may converge to a
different point from the other two; in this case a short
and straight central suture connects both pairs. No other
plant, recent or fossil, is known to present a like feature.

[1] Sigillarian stems with more than four principal root-masses are
known, and it is probable that the mode of division varied according to
the species.

The main roots soon divide into smaller branches, and
these into yet smaller, but the ultimate divisions are of
considerable size, when compared with the delicate filaments into which the roots of one of our existing forest-
trees break up. The Stigmarian roots are covered with
round scars arranged quincuncially, and these give attach-
ments to the rootlets, each of which is traversed by a

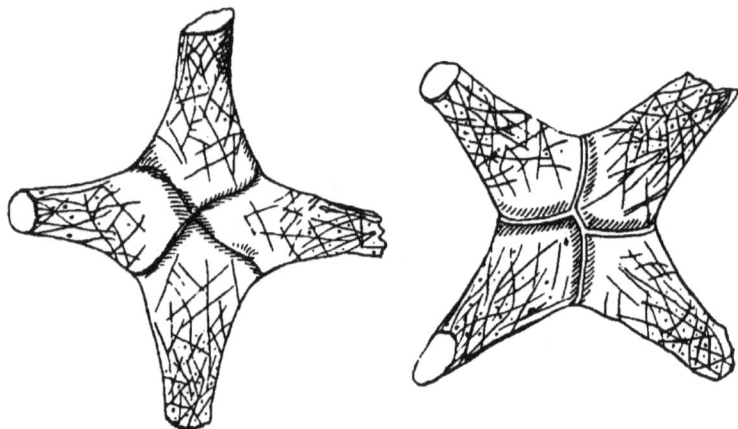

FIG. 7. Main roots of Stigmaria (under side) showing sutures.

bundle of vessels derived from the woody zone. The
rootlets are united continuously to the outermost cortical
layer by a bottle-shaped base (fig. 9).

The most distinctive outward features of a Lepidodendron stem are the repeated forking (dichotomy) of the
branches and the quincuncial, lozenge-shaped leaf-scars.
Long, narrow leaves densely clothe the young twigs.

The internal structure varies not inconsiderably
according to the species. We may take Lepidodendron
selaginoides as a tolerably central type connecting the
extreme modifications. When the stem of this species is
examined by means of transparent sections a 'medullary
axis' is found to occupy the centre. This consists of
thick-walled 'barred' vessels, not unlike those so distin-

guished in Calamites; barred cells, not set in vertical columns, and retaining unabsorbed their original cell-walls; and thin-walled parenchyma or cellular tissue. These are more or less intermingled, but many of the vessels are aggregated so as to constitute an outer invest-ment (the vascular medullary cylinder) to a central vasculo-cellular mass. From this medullary axis are derived the vascular leaf-bundles, which traverse the woody zone and the bark to enter the leaves. External to the medullary axis a second vascular zone or cylinder is found. This consists of barred vessels arranged regularly in vertical radiating lamellæ, and increasing outwards in size as well as in number of series. The varying thickness of this cylinder, the increase of its constituent vessels from one to perhaps some hundreds in a row, and their progres-sive delicacy, fragility, and larger size as we pass outwards, sufficiently indicate the exogenous character of this zone; i.e. its growth by successive additions of new vessels to its exterior. The woody cylinder occupies a tract which was primitively cellular. It may be supposed that the original cellular tissue was compressed by growing wedges of wood into vertical plates. Such vertical plates radiating like the series of vessels, and composed of simple cellular tissue, are regularly found in transverse sections of the Lepidodendron stem, and are distinguished as the medullary rays. Whatever the mode of their original formation, they increase regularly and exogenously, like the intervening constituent wedges of the woody cylinder.

The bark was thick and divisible into layers. The innermost of these consisted, like the cambium-layer of a recent exogenous tree, of a dense and ever-renewed layer of minute cells, capable of conversion into more complex tissues, and yielding thus a perpetual accession of fresh material to the layers within and without. The middle layer of the bark was made up of a slightly coarser cellular tissue, gradually replaced outwards by vertically elongate and thick-walled cells (the 'bast-layer.') To the outer surface of the bast-layer another cellular invest- ment, the 'epidermis' or outer bark, was closely applied.

Lepidodendron Harcourtii is distinguished by the ab- sence of any exogenous cylinder surrounding the vascular zone formed by the confluent leaf-bundles. Other types again depart from Lepidodendron selaginoides in an oppo- site direction, and are characterised by the high develop- ment of the exogenous cylinder, the numerous vessels which compose each radiating lamella, and their very definite arrangement.

A parallel series might be arranged by comparing, not selected species, but successive stages of the same species. A very young Lepidodendron twig consists of ordinary cellular tissue, with a single central bundle of barred vessels enclosing a rudimentary pith. As the stem increases, the bundle of vessels enlarges to form the ordinary vascular medullary cylinder, in which the vessels are not arranged in radiating series. This cylinder sup- plies the vascular bundles which go to the leaves. About the same time the outer cellular investment becomes differentiated into the inner and outer layers of the bark, to which a third and intermediate layer is soon added. When this stage is completed the stem may be held to correspond with that of Lepidodendron Harcourtii.

An exogenous growth of vessels now takes place in a cambium-layer which surrounds the vascular medullary cylinder. The new vessels are added in radiating laminæ, and separated from each other by medullary rays. We thus reach the stage corresponding to Lepidodendron selaginoides, and by continued development the yet more complex stems are produced. It is the exogenous

cylinder which especially distinguishes the adult states of
the more highly organised species—first, from young
twigs, secondly, from such forms as Lepidodendron Harcourtii, and lastly from the recent Lycopods.

It will be observed that the exogenous zone is not in
direct communication with the leaves, though continuous

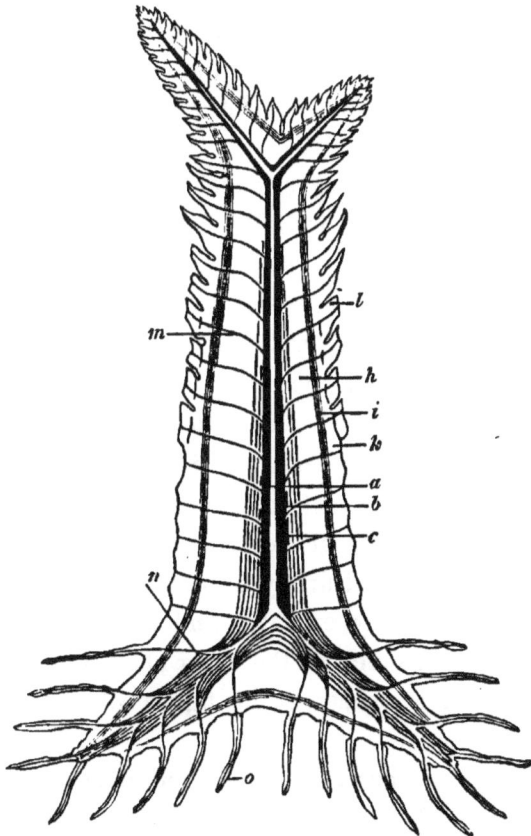

FIG. 8.—Diagrammatic vertical section of Lepidodendron. *a*, pith ; *b*,
medullary vascular cylinder ; *c*, exogenous woody zone ; *h*, *i*, *k*, layers
of bark ; *l*, leaf-stalks and leaves ; *m*, vascular bundles of leaves ; *n*,
vascular bundles of roots ; *o*, rootlet. From Williamson ('Phil. Trans.
vol. clxii. pl. ii. 1872).

with the vascular bundles of the roots through the exogenous cylinder shortly to be described in Stigmaria.   On
the other hand, the vascular medullary cylinder of the

stem is in close relation with the leaves by means of the vascular bundles of which it is made up, but it does not extend into the roots, nor has it any continuity with the exogenous zone. Food-material taken up by the roots would therefore ascend freely into the exogenous zone of the stem, and there find itself cut off from the leaves; similarly food-material formed in the leaves would descend into the vascular medullary cylinder, but be unable to pass into the exogenous zone or the roots. The channel of communication between the two tissue-systems was provided by the medullary rays, which are in the closest relation to both, passing radially from one to the other; or else the fluids passed laterally by osmosis (diffusion) through the walls of the several vessels.

It appears from the state of preservation of fossil Lepidodendra that the woody cylinder was particularly liable to become detached, owing to the decay of the delicate inner and middle bark. It is rarely found intact. Most of the fossils are casts of the exterior of the epidermis, or of the inner surface of the bast-layer. It seems that these adjacent and firmly united layers often resisted decay after the woody cylinder and inner layers of bark were wholly gone, much as the outer shell of a tree-trunk in a tropical forest preserves its figure when the central wood is decomposed.

The vascular bundles briefly noticed above pass obliquely outwards and upwards from the medullary axis to the leaves. They consist of small collections of vessels surrounded by cellular tissue derived from the inner bark. Apertures (the 'lenticular spaces') are left in the exogenous or radiating woody cylinder and the bast-layer for their passage. The lenticular spaces, like the leaves, are disposed quincuncially, and the quincuncial bosses which mark their position are characteristic features of the internal casts of the bast-layer (formerly named Knorria, on the supposition that they were independent stems).

The differences of internal structure between the stem of a Lepidodendroid tree and the root (Stigmaria) require a few words of explanation. Stigmaria has the central pith, the radiating and exogenous laminæ of the woody zone, the medullary rays, and the thick bark of the Lepidodendron stem. But the inner vascular cylinder, forming part of the medullary axis and sending off vascular bundles to the leaves, is wanting in the root. It is re-

Fig. 9. Restoration of part of Stigmaria, with attached rootlets. From Williamson ('Phil. Trans.' vol. 162, pt. 1, 1872).

placed by vascular bundles derived from the vascular laminæ of the woody zone, which pass outwards in tongue-like processes through meshes in the wood and bark. The vessels in each bundle gradually diminish in number towards the outside of the stem, and but few actually pass out into the rootlet. The cellular sheath which invests the vascular bundle of the rootlet unites by its flask-shaped base with the thick epidermis or hypoderm of the main root.

Comparing the Lepidodendron stem just described with one of our recent forest-trees we may notice, as

points common to both, the central pith; the woody zone,
exogenous, or increasing by external addition, and parted
into wedges by radiating vertical plates of cellular tissue
(the medullary rays); the bark divided into layers; and
the vascular bundles given off to the leaves. Points of
difference are no less easy to discover. The woody zone
of Lepidodendron contains none of the spindle-shaped,
thick-walled cells, which characterise the wood of our
common trees, but in their place 'barred' vessels, large
as compared with woody cells, and regularly placed in
vertical radiating planes. Lepidodendron, moreover,
shows no rings of annual growth. This is, as we have
seen above, a common feature in the Coal Measure plants
of Britain, and may indicate the effect of some external
cause, such as an equable climate, in which vegetation
underwent no periodical check.

A recent Lycopodium stem seems at first sight extremely different from a Lepidodendron. It is herbaceous,
green, and flexible, attaining a length of a very few feet,
and is usually prostrate. When cut across one or more
vascular bundles are seen enclosed in a mass of cellular
tissue, which extends between and around them from the
centre to the epidermis. Each vascular bundle is encased
in a special cellular sheath, with the addition, in Selaginella, of surrounding air-spaces, and each bundle gives off
fasciculi of vessels to the leaves. To convert such a stem
into that of Lepidodendron, the detached vascular bundles
must unite into a ring, so as to enclose a central pith-
cavity; they must receive an entirely new external and
exogenous cylinder (it will be remembered that this is
wanting in Lepidodendron Harcourtii); and the cellular
tract outside must arrange itself into distinct layers
around the now complete woody zone. Such changes
seem almost to amount to entire reconstruction, yet the

Lepidodendron and Lycopodium are after all near of kin, and even their stems have a common fundamental structure. The difference is that between the arborescent and

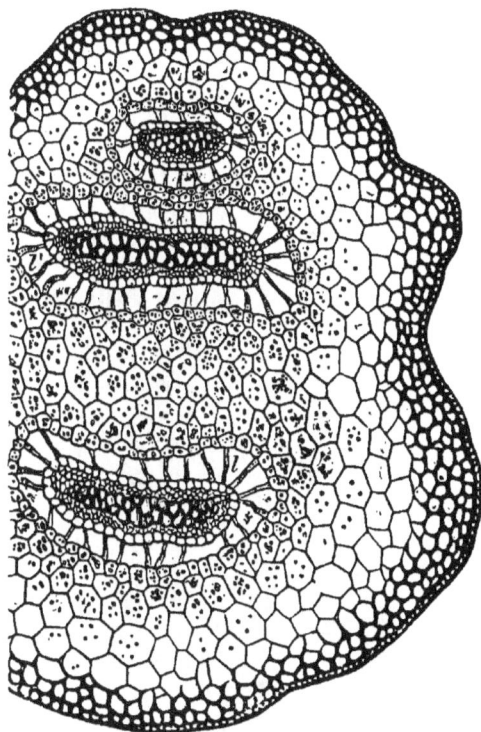

FIG. 10. Transverse section of stem of Selaginella (× 150).   From Sachs'
' Botany.'

herbaceous forms of the same type; it indicates the extent to which physiological adaptation may mask real affinity.

The fruit of Lepidodendron (often called Lepidostrobus) is more important than any other part of the plant for systematic purposes. It has a somewhat cylindrical form, but is rounded above and below. Externally it is densely covered with overlapping scales, which are merely leaves slightly modified. The internal structure of a Lepidostrobus will be best understood by comparison with the the corresponding part of a recent club-moss.

In Lycopodium certain leaves support at their bases sporangia or spore-cases. These fertile leaves usually differ in form from the ordinary foliage; though in one

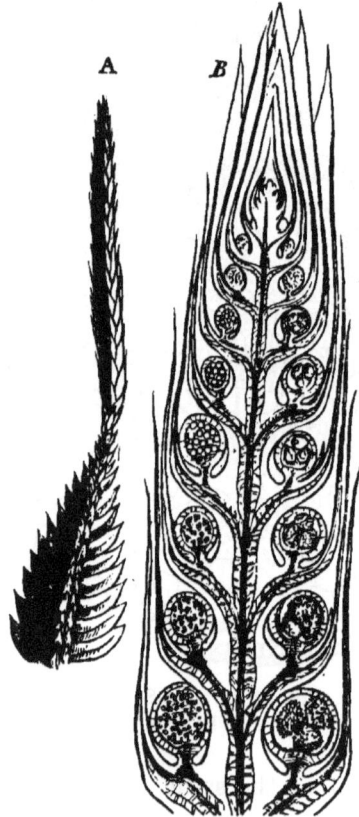

FIG. 11. Selaginella. A, fertile branch (one-half natural size). B, apex in longitudinal section, bearing microspores on the left, and macrospores on the right (magnified). From Sachs' 'Botany.'

English species (Lycopodium Selago) they are nearly uniform with them, and the upper part of the spike, which bears sporangia, is in this species hardly distinguishable in outward appearance from the rest. In our other Lycopodiums the fertile leaves are clustered together

into separate cones, which differ from the sterile branches
in shape and size (resembling thus the fossil Lepidostrobus), and also in colour. If one of these fertile Lycopodium cones be cut through vertically, the sporangia are
seen nestling in the axils of the leaves, one at the base of

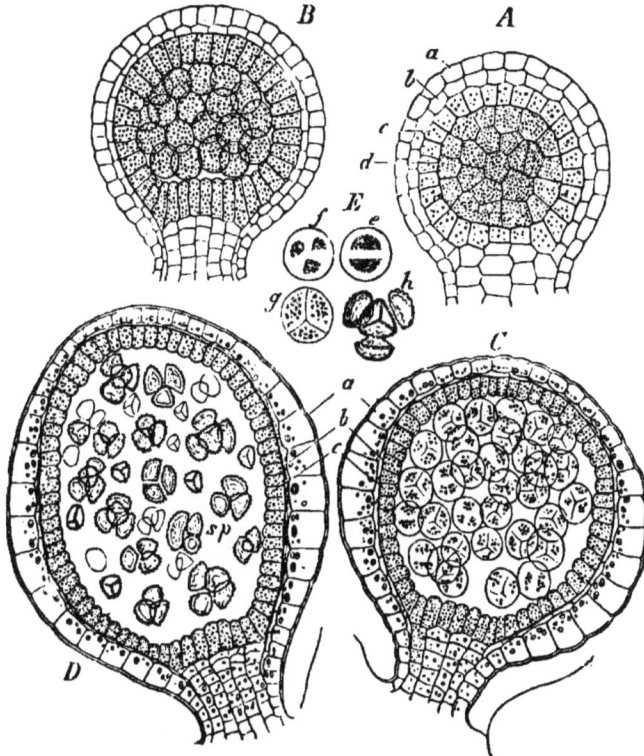

FIG. 12. Development of sporangia and spores of Selaginella, in order of
    the letters A—D. A, B serve for all the sporangia; C, D for micro-
    sporangia only. E, division of mother-cells of microspores; *h*, four
    nearly ripe spores. (Magnified.) From Sachs' ' Botany.'

each. The contents of a sporangium, originally a cellular
mass, (fig. 12, A) arrange themselves gradually into an
outer cellular wall and a cluster of inner cells, the ' mother
cells.' These give rise by division to spores, four to each
mother-cell. In Selaginella, a genus of Lycopodiaceæ
abundant in some tropical countries, but represented by a

single and not very common species in the British Islands,
the spores are of two kinds, and the capsules in which
they are lodged differ also when mature. In Lycopodium
but one kind of spore appears to exist. Many of the
Carboniferous club-mosses are of the Selaginella type,
and we shall therefore take this form for express comparison. Selaginella, then, has sporangia of two kinds;
in the one kind (macrospongia, which occupy the lower
part of the cone) a single mother-cell grows at the expense
of the rest, and enlarges so as to fill a great part of the
cavity. It ultimately breaks up into four masses by the

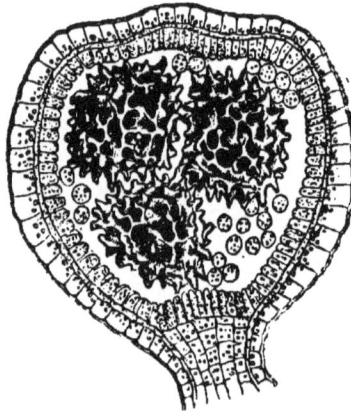

FIG. 13. A nearly ripe macrosporangium of Selaginella; the fourth spore
       is not seen (× 100). From Sachs' 'Botany.'

subdivision of the cell-protoplasm. Each of the resulting
quarters (macrospores) is a three-sided pyramid with a
rounded base, and these ultimately break away from each
other and lie loose in the ripe sporangium, together sometimes with the much smaller undeveloped mother-cells
(fig. 13). In the other kind of sporangium (microsporangium),
the growth of the mother-cells is more equal. A number
of comparatively small microspores (fig. 12, D) similar in
form to macrospores, fill the ripe microsporangium. The

subsequent processes of growth are immediately concerned with reproduction. The microspore, after liberation from its capsule, undergoes division into cells, most of which give rise to antherozoids. These are threads twisted into a corkscrew shape, and furnished at one end with two long hairs or cilia, which produce a slow spiral motion for some time after the escape of the ripe antherozoid from its parent-cell. Similar particles, endowed for a short time with automatic motion, occur in male animals; they correspond, at all events physiologically, with the contents of the pollen-tubes of a flowering plant, while the microspores themselves are comparable to pollen-grains. The macrospore, too, escapes from its capsule and enjoys a short term of independent life, producing by internal division many small cells, which remain attached and form tissue. The growth of these contained cells at length bursts the envelope of the macrospore and a minute scale (the ' prothallium ') is protruded. The prothallium pushes tiny rootlets into the ground, while minute crater-like cavities ('archegonia') open upon its upper surface. No archegonium is developed in flowering plants, but the embryo-sac, or the germinal vesicle of an animal ovum, plays nearly the same part. A single archegonium is first developed, and lies ready to receive an antherozoid. If any wriggling filament reach the microscopic orifice, it enters, and quickly sets up active growth in the potential embryo, at first a mere rounded cell lying in the bottom of the archegonium, or receptive cavity of the prothallium. If no antherozoid should be at hand, the first archegonium soon withers away, but others rise up, one after another, so as to give fresh opportunities of fertilisation. The embryo, once fertilised, begins its individual career. It forms fresh cells, and these arrange themselves into organised tissues. A

small plant springs from the prothallium, whose part
is now played out. The young Selaginella rises out of
the earth, grows bigger and bigger, turns green (showing
thereby that it has begun to form food-material for
itself), and finally reproduces the parent form, upon which
the sporangia were originally developed.

To form a clear notion of this intricate chain of
growths and processes, the prothallium may be supposed
to be physiologically, though not morphologically, comparable to a detached flower. In ferns, and perhaps in
Lycopodium, but one kind of spore is produced, and this
bears both male and female organs—stamens and pistils,
or as they are called in this case antheridia (antherozoid-
cells) and archegonia. In Selaginella two kinds of
spores, comparable to two kinds of flower-buds, such as
are called diclinous among flowering plants, are produced ;
the one ultimately developes antherozoids or fertilising
agents, the other archegonia, in which embryos, at first
unfertilised, lie hid. The familiar forms known as ferns
and club-mosses constitute the Asexual Generation of the
plant; in this state it may bud off fresh individuals, but
it never directly produces fertilised embryos. The prothallium, on the contrary, the inconspicuous terrestrial
or subterranean leaf-scale, is the Sexual Generation,
never increasing by budding, but producing the fertilised
embryo by a true reproductive act.

Reverting now to Lepidostrobus (the fruit-cone of
Lepidodendron), we shall find the closest correspondence
with the fertile spike of a Selaginella. The spore-bearing
leaves are attached in a dense spiral to the central axis, and
in such manner that their outer ends, curving upwards,
form an imbricate diagonal pattern, very like the quincuncial ornament of the stem. Each scale bears a sporangium upon its upper surface. In some cases certainly,

and possibly in all, the basal sporangia of the cone produced macrospores, and the apical sporangia microspores.
The spores themselves have been revealed in sections of
the cone. When examined by a low power of the microscope they are found either to cluster in fours, or to
indicate by their figure, which is that of a three-sided
pyramid, that they have broken away from such a combination.[1] We can desire no better proof of the lycopodiaceous nature of Lepidostrobus than is afforded by these
points of agreement. The association of Lepidostrobus
with Lepidodendron, again, is vouched for by unusually
clear evidence. Not only is the central axis of the cone
identical in structure with the ordinary leaf-bearing
branches, but the cone itself has often been found attached
to a quite unmistakable bough of Lepidodendron. It remains to point out the correspondence in form between the
macrospores of Lepidodendron and those found in such
countless multitudes in many coals (supra, p. 20). The
size, about $\frac{1}{20}$ of an inch in each case, is in itself a possible means of identification, but when it is added that the
compressed macrospores of the coal exhibit a 'three-rayed
mark,' which plainly represents the meeting faces of the
three-sided pyramid of the Lepidostrobus spore, possibility
may be said to have become certainty.

The lengthy examination of Calamites and Lepidodendron into which I have entered leaves but little space
for other Coal Plants. There are many ferns, and some
of these are clearly proved to have been tree-ferns. It
rarely happens that the fructification, consisting of clusters of spore-cases on the back of the frond, is sufficiently
preserved to allow of comparison with recent species.
This is the more unfortunate since the botanical affinity
of a fern is mainly determined by its fructification. Form

[1] Combinations of three have been said to occur.

of frond counts for nothing, venation for little. Dr.
Hooker figures side by side four ferns, practically identical in outline and venation, but differing totally in the
sori, or clusters of spores, and thereby shown to belong to
four perfectly distinct genera.[1] Nine-tenths of the names
of Carboniferous ferns rest upon mere comparison of outline, and are, like too many other technicalities of palæontology, altogether worthless.

One coniferous tree (Dadoxylon) is not uncommon in
the Coal Measures of Britain. Its wood is generally
similar to that of many recent Conifers, and the microscopic 'bordered discs' of the wood-cells, characteristic
of the order, though not absolutely confined to it,[2] have
been more than once made out in this fossil tree. The
transversely corrugated cast of the large pith-cavity
(Sternbergia) has been found enclosed in stems of Dadoxylon. It has already been remarked (p. 79) that Dadoxylon rarely exhibits rings of yearly growth.

M. Grand' Eury is disposed to assign Dadoxylon to
the Cycadeous genus Cordaites, but this decision should
be received with hesitation. Cordaites exhibits a somewhat low, unbranched stem, crowned by a cluster of
sword-shaped, parallel-veined leaves. Slender terminal
spikes support small scaly ovules in the axils of bracts,
and represent the female inflorescence. Male flowers
with anthers are borne upon separate spikes. It is
affirmed that the fossil flowers named Antholithes and
the fruits known as Cardiocarpon have belonged to Cordaites, but several previous identifications of these parts
with other forms of vegetation ought to render us cautious.

[1] *Mem. Geol. Survey*, vol. ii. pt. 2, p. 408.
[2] Such discs occur also in Cycads, Winter-barks, Bucklandia and
Sedgwickia.

M. Grand 'Eury thinks that some coals of Central France are entirely composed of the remains of Cordaites.[1]

Diatoms are said by Count Castracane to have been discovered in coal by ignition and boiling in strong acids. Microscopic examination of the indestructible refuse yields the siliceous valves of these Algæ, which are described as agreeing precisely with recent forms.[2] Confirmation of these statements is much to be desired.

The mycelium of a fungus is described by Mr. Carruthers as ramifying within the vascular tissue of a Lepidodendron,[3] and other supposed Carboniferous fungi are on record, but these determinations are not beyond doubt. Considering the perishable nature of fungi, we can hardly hope for conclusive evidence of their presence in ancient rocks.

The existence in the Coal Measures of flowering plants higher than Coniferæ has been affirmed, but here again we wait for further proof.

Many other plant-remains—foliage, stems of unknown trees, gymnospermous seeds—are known to exist, and will no doubt ultimately enrich our catalogues of Coal Measure fossils. It would be useless to discuss here forms whose affinities are not even approximately determined, but the reader will do well to bear in mind that a mass of undigested material is extant, the symbol of much more which is as yet inaccessible to the student.

Tabulating the Carboniferous flora according to our present (most imperfect) information, and setting it side by side with the existing vegetable kingdom we get the following summary view:

---

[1] Grand 'Eury, Flore Carbonifère du département de la Loire, *Mem. Acad. Sci.*, tom. xxiv. 1877.

[2] Castracane, *Compt. Rend.*, tom. lxxix. p. 52. 1874.

[3] *Pres. Address to Geologists' Assoc.* 1875.

	RECENT	CARBONIFEROUS
CRYPTOGAMIA	Algæ	Diatomaceæ
	Fungi	?
	Hepaticæ	
	Mosses	
	Ferns	*Pecopteris, &c.*
	Equisetaceæ	*Calamopitus, Calamites.*
	Ophioglossaceæ	*Myelopteris (Palmacites* of Corda)
	Rhizocarpeæ	
	Lycopodiaceæ	*Lepidodendron, Sigillaria, &c.*
PHANEROGAMIA	Gymnospermæ	*Dadoxylon, Cycadeæ.*
	Monocotyledones	*Pothocites?*
	Dicotyledones	

It is obvious that the absence of particular classes and orders from a fossil flora is no proof that they did not once exist. Special external conditions, special features of structure or composition favour the preservation of some and the complete dissolution of others. It is however to be observed that the Carboniferous flora, imperfectly as we know it, already includes examples of nearly all the great divisions of existing plant-life which we might expect to be capable of recognition in a fossil state. Possibly all the palæozoic species are now replaced by new ones, possibly all the genera. Of the larger groups—families and orders—very many have passed away, very many have come in. But when we turn to the primary divisions established by botanists, we seem to recognise the fundamental identity of the existing vegetable kingdom with that of the Coal Measures. The most striking deficiencies of the ancient flora are those of the Mosses and the Dicotyledons (to which most of our common trees and herbs belong); the presence or absence of Monocotyledons cannot as yet be affirmed. It is not less important to observe that the Carboniferous flora, while furnishing many curious modifications of existing types—of which perhaps the most remarkable are the exogenous club-mosses and horse-tails—adds no new class or primary

division to those founded upon recent plants. The vegetable kingdom of to-day is no new creation, it is in its
broad features at least as ancient as the Carboniferous
rocks. Nor are the Carboniferous plants strange and
anomalous to the student of living plants; each is united
to some surviving form by ties of resemblance which
every fresh investigation strengthens.

CHAPTER IV.

ANIMALS OF THE COAL MEASURES.

FROM the animal forms which have been brought to
light by searching the Coal Measures we shall select the
vertebrates for a chief share of attention. Of these the
highest in the zoological scale are Amphibia. It may be
worth while to remind the reader of some few characters
of this class. Amphibia have much in common with
Fishes, with which they are very closely associated in the
systems of modern naturalists. The points of resemblance,
numerous and important as they are, relate, however, for
the most part, to deep-seated organs, familiar only to the
anatomist. One common feature is apparent without dis-
section; both Fishes and Amphibia breathe by gills during
part or the whole of life. But few constant differences can
be mentioned. The lungs, always present in adult Amphibia, whether the gills are retained or not, do not furnish a
diagnostic character, for the lung is present as a swim-
bladder in many fishes, and may even in certain species, as
will be seen further on, discharge temporarily the respiratory function. It is more important to notice that Amphibia, though they may have a median fin upon back and tail
like fishes, never possess fin-rays; and that the limbs of
Amphibia, which correspond to the paired pectoral and
ventral fins of fishes, agree in the general arrangement of
their bony elements with those of higher vertebrates,
while they differ in this respect from any of the forms of

supporting skeleton found in pectoral or ventral fins. The occipital condyles, or pivots of suspension of the skull upon the vertebral column, are always double in Amphibia—a character not available for the separation of Amphibia from fishes, but of great practical value, especially to the palæontologist, in discriminating them from reptiles, in which the condyle is invariably single.

It is not improbable that more than one order of Amphibia is represented by fossil remains in the Coal Measures, but the best investigated species belong to the extinct order known as Labyrinthodonts, and of these we shall proceed to give some account.

Labyrinthodonts were first recognised in the Triassic rocks of Germany, where fifty years ago some obscure fragments and afterwards a nearly perfect skull of the large form named Mastodonsaurus were brought to Dr. Jaeger of Stuttgart for examination. It was far from easy to interpret these fossils satisfactorily. The large, broad skull, with powerful conical teeth, outwardly similar to those of Crocodiles or Ichthyosauri, suggested Saurian affinities ; on the other hand the paired occipital condyles, which formed a prominent feature of the back of the skull were, according to an unbroken experience, wanting in all true reptiles, recent or fossil. The chief weight of opinion among German naturalists inclined to the view that Mastodonsaurus was really a Saurian, though Jaeger himself laid stress upon the double condyle. Fresh discoveries soon followed. Von Meyer, Münster, and Braun described several new species nearly allied to Mastodonsaurus, and in 1838 Von Meyer called attention to the peculiarities of the teeth, which have since been more minutely examined. Meanwhile in England bones of animals very similar to Mastodonsaurus had been discovered in the New Red Sandstone of Warwick as early as 1823, but it was not till

1841, or shortly before, that they were carefully examined
and compared.   Prof. Owen, to whom the English specimens were submitted, described their structure with a
sagacity and a command of anatomical knowledge which
has placed his memoir high among contributions to
palæontology.   He made fully known the remarkable
structure of the teeth, already noticed in Mastodonsaurus
by Von Meyer, correlated perplexing fragments of bone
with fair success, and advanced several ingenious and on
the whole instructive propositions respecting the outward
form and zoological place of the group, to which he gave
the name of Labyrinthodonts, from the complicated fold-
ing of the wall of the tooth.   Nearly forty years, rich in
palæontological discovery, have elapsed since the publication of this memoir, and it is now easy to point out errors
and defects ; none the less, Prof. Owen's paper 'On the
Labyrinthodons of Wirtemberg and Warwickshire ' will
long be remembered as a signal example of the successful
interpretation of fossil fragments by the light of recent
comparative anatomy.

Prof. Owen was clear that these extinct animals were
Amphibia, or as they were then called according to
Cuvier's nomenclature, Batrachia.   In the reconstruction
of the Labyrinthodonts, he introduced comparisons with
several Amphibian types, but finally retained a more or
less distinct impression that in outward form at least they
resembled the frogs and toads, which form in modern
zoology the restricted group of Batrachia (a *section* only
of the class Amphibia, and not to be mistaken for the
Batrachia of Cuvier or Owen's memoir).   This view was
suggested or confirmed by the following facts.   First,
there had been found in the same New Red Sandstone of
Warwickshire limb-bones, which exhibited a remarkable
inequality of fore and hind legs.   It had also been ascer-

tained that in the New Red Sandstone both of Great
Britain and Germany there occurred foot-prints of an
unknown animal provisionally named Cheirotherium, in
which the impression of the hind-foot was greatly larger
than that of the fore-foot. Dr. Kaup had conjectured
that this might be the track of a marsupial. Again,
distinct sizes of foot-prints, possibly due to distinct species
of animals, had been observed; while the bones, which
were referred to several species of Labyrinthodonts,
exhibited a comparable difference of dimensions. Bearing
in mind the amphibian nature of the Labyrinthodonts, of
which there was adequate proof, it seemed natural to
conclude, though this was done with suitable, hesitation,
that the limb-bones found in Warwickshire belonged to
the same kind of animal as the fragments of Labyrinthodont skulls; that the Cheirotherian foot-prints were
made by Labyrinthodonts; and that they 'were the foot-
prints of a reptile,[1] and not of a marsupial or other
mammal, and that this reptile most probably belonged to
that family of the class which includes the Frog and
other anourous (tail-less) Batrachia, which offer a similar
disproportion between the fore and hind-legs.' To
illustrate this supposition a diagram was prepared, in
which the now familiar restoration of Labyrinthodon as a
toad-like animal was displayed.

It will be admitted that the hypothesis which thus
connected several independent facts into one consistent
view was at least ingenious and plausible, but we must
next consider to what extent it agrees with the fuller
knowledge since acquired. Little direct information is
even now accessible respecting the limbs or general pro-

---

[1] It should be remarked that at the time of Prof. Owen's memoir the
term 'reptile' was universally understood to include the Amphibia. At
present, one of the most definite of zoological lines is drawn between
Amphibia and Reptilia

portions of either Mastodonsaurus or Labyrinthodon, but
the other and more perfect examples of the order, which

FIG. 14. *Proteus,* a recent Amphibian from subterranean caverns
of Carniola and Istria.

FIG. 15. *Menobranchus,* a recent
Amphibian from North America.

have of late years been discovered in considerable numbers,

resemble not the anourous but the tailed Amphibia.
They have short, feeble limbs, adapted for motion in an
element which is capable by its own density of supporting the weight of the body, and there is no important
inequality of fore and hind legs. It has also been
ascertained that most and perhaps all the Labyrinthodonts,
certainly the species in question, were provided with a
defensive armour; that the bones of the skull were
greatly thickened; and that the size and weight of the
large Triassic species vastly exceeded those of any existing amphibian. Even forty years ago the bearing of these
last mentioned considerations might have been perceived,
though the facts were less perfectly established than at
present. Even then it was open to the naturalist to
mark the discrepancy between a structure at once heavy
and weak, and the agility implied by excessively developed
hind-limbs. The frog or toad owes its form and especially
its extraordinary length of hind-legs to its leaping habits.
The Amphibia which pass their whole life in the water, or
move sluggishly on land, retain the tail, and have feeble
and generally approximately equal extremities. But what
can the toad-like form mean in the case of a Labyrinthodont? An animal several feet in length, laden with bony
armour, provided with a broad skull (in Mastodonsaurus
thirty inches long), and with massive, conical teeth in
double rows, bearing this skull upon a slender and ill-knit
spinal column made up of biconcave vertebræ—such an
animal could not leap a yard in air, and could not fall a
yard without fracture and dislocation. The peculiar
mode of life of the frog or toad, and the peculiar structure
of limbs implied thereby, are totally inapplicable to the
Labyrinthodont—a creature which whether for leaping or
alighting would be at a disadvantage even when compared with Crocodiles or Turtles.

Such considerations as these justify us in refusing to accept either the pictorial or the verbal restoration of Labyrinthodon, but a mistake is never satisfactorily disposed of until it is explained, and we may usefully endeavour to find out the weak place in Prof. Owen's reasoning. It lies perhaps in the tacit assumption that very few species of vertebrate animals higher than fishes inhabited the earth in Triassic days, and that correspondence in size or association in the rocks is therefore a fair presumption of identity. This argument, when stated in words, is evidently far from cogent. In all probability the great majority of the large animals of this period remain wholly unknown to us, but it is at least ascertained that Lacertilia, Plesiosauria, Ichthyosauria, Crocodiles, and Dinosauria lived in Europe during Triassic times. There were also small mammals, and very likely birds too. It is by no means improbable that the limb-bones found in Warwickshire, or some of them, belonged to Dinosauria, and the Cheirotherian foot-prints may be Dinosaurian also. The settlement of these points cannot be attempted here; it is sufficient for our present purpose to know that there are many alternatives, if the Labyrinthodont character of any New Red Sandstone fossil be rejected.

Under these circumstances the inherent improbability of a tail-less Labyrinthodont with excessively prolonged hind legs must be allowed its full weight. We shall do well to banish the old wood-cut, at which more than one generation of geologists has gazed with satisfaction, and to replace it by another figure—that of a tailed, aquatic animal, feeble on land, but swift and rapacious in the water, a Crocodile in outward form and habits, though Amphibian in its derivation and internal structure.

The Triassic Labyrinthodonts exhibit more conspicuously than those of earlier date the peculiar tooth-

structure which has given a name to the order. From
Archegosaurus we learn that the tooth is first formed as
a small hollow cone of enamel with two cutting edges.
This, the true crown of the tooth, retains its original size
and structure until it disappears by wear or fracture. It
does not, however, remain directly attached, as at first,
to the dentigerous bone, but is gradually elevated by the

FIG. 16.   Section of tooth of Mastodonsaurus (magnified).

growth of a continually enlarging base. This base, which
is often the only part of the tooth remaining, has the form
of a hollow cone of tooth-substance (dentine), coated
thinly with enamel, and enclosing a pulp-cavity. Its wall
becomes longitudinally and radiately folded, as if frilled,
and in the Triassic species some or nearly all of the plaits
may be again folded tangentially to the surface of the
tooth. The folds are ordinarily close together, and leave

only linear spaces between. In this way the thickness of the tooth-wall is greatly increased, and the central cavity much encroached upon. The enamel may extend itself between the dentine as a thin undulating layer. A cross section of such a tooth as has been described shows a set of sinuous and branched passages, which originally communicated with the exterior, but are now obliterated or occupied by enamel, and a corresponding set (divided from the first by a layer of dentine) which represent extensions

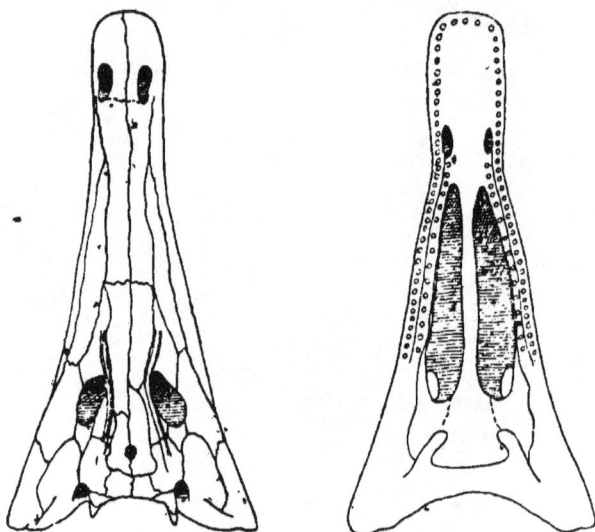

FIG. 17. Upper and under surfaces of skull of Archegosaurus.

of the pulp-cavity. In some of the Carboniferous Labyrinthodonts there are no secondary folds of dentine, and the degree of convolution varies from a simple undulation to a complex and far from regular labyrinth. As a rule the complexity increases with the size of the species. The teeth are lodged in shallow depressions, which take the form of the base, and are often marked by radiating ridges corresponding with the folds of dentine.

The existence of Labyrinthodonts in the Coal Measures

CHAP. IV.   ANIMALS OF THE COAL MEASURES.   119

was first rendered probable by the discovery, in 1842, of a
fossil Amphibian at Münster-appel in Rhenish Bavaria.
This new form was described by Von Meyer in the following
year, under the name of *Apateon pedestris*.   It is a small
and in many respects obscure specimen, but there is now
no doubt at any rate as to its Amphibian character.   The
bones of the limbs, though imperfect, are sufficient to
distinguish it from any fish.   In 1847 several examples
of the Labyrinthodont now known as *Archegosaurus Decheni*
were discovered in nodules of clay-ironstone at Lebach
in the Saarbrück coal-field by Von Dechen.[1]   Precisely
similar fossils had been discovered long before in the
same locality, but their real nature had escaped detection.
A skull, catalogued in 1777 as part of Pasquay's collection
(now in Stuttgart), had long passed for a fossil fish, and
had been named *Pygopterus lucius* by Agassiz on this sup-

[1] The deposits from which these fossils come are now often placed by
foreign geologists in the Permian.   This is done mainly or altogether on
the strength of what are called characteristic Permian fossils.   When the
case is looked into, it is not very convincing.   The Carboniferous flora is
admitted to pass upwards into the beds in question.   The fishes quoted
as characteristic are Xenacanthus (Diplodus), Acanthodes, Ctenoptychius,
Palæoniscus, Amblypterus, Pygopterus, and Platysomus.   Others add
Ctenacanthus and Pleuracanthus.   All the genera are found in the English
Coal Measures, and the well-determined species are either Carboniferous
or peculiar to the beds in question.   It is the same with the invertebrates;
all are Carboniferous or peculiar.   No Mollusca are mentioned.   We are
not disposed to insist upon the division between Carboniferous and Permian (though locally useful) as of world-wide application.   The mineral
difference between the English Coal Measures and the Magnesian Limestone, the frequent unconformity, the difference of fossils, may be very
largely explained by merely local events, such as change from a fluviatile
and terrestrial surface to a land-locked sea, more and more largely charged
with mineral salts, or interruption of deposit accompanied by waste of
previously formed rocks.   We need not be surprised to find elsewhere
stratigraphical continuity with gradual replacement of species, and indeed passage-beds between the two formations are already described.
Reasons of this kind might be given for abolishing the boundary altogether, but we demur to transferring it to another place, merely because
certain common fossils of the English Coal Measures have been called
exquisit Permische Thier-reste.

position.   It is in reality a skull of Archegosaurus, but so
fish-like in general appearance that we cannot charge the
naturalists who failed to perceive its true zoological position with either dulness or negligence.   Hundreds of
examples of Archegosaurus, some fairly complete, are now
preserved in the museums of Germany.   These have been
carefully studied and described, especially by Hermann
von Meyer, who has also figured many of them with his
own hand.

Certain details of the organisation of Archegosaurus
have materially contributed to our knowledge of the
Labyrinthodont order, and have helped to place its
zoological rank beyond dispute.   In this genus the first
traces of branchial arches were discovered.   Some specimens exhibit scattered ossicles in the region of the throat,
which appear to have been disposed in three or more rows,
and doubtless carried respiratory filaments.   In subsequently discovered genera the branchial arches have been
still more plainly seen, and all question as to the class to
which the group belongs is thereby set at rest.   No
reptile, recent or fossil, exhibits more than a trace of bony
gill-arches in the form of a hinder cornu or horn to the
hyoid bone.   We may therefore rank the Labyrinthodonts without misgiving or hesitation   as   Amphibia.
These ossicles have been found only in young examples,
and as they do not, so far as is known, increase with age,
it is inferred that the gills were not persistent.   Again,
it was in Archegosaurus that the remarkable ventral
armour, now believed to characterise all true Labyrinthodonts, was first fairly exposed to view (fig. 18).   This
consists of three thoracic plates and many rows of small
abdominal scutes.   The central thoracic plate is diamond-
shaped and of relatively large size; it is flanked and
partly overlapped by lateral plates of triangular form;

all three are covered with a radiating sculpture, not unlike that of the bony scutes of some Crocodiles. Behind the thoracic plates are rows of abdominal scutes, which converge backwards towards the middle line; the pattern is reversed in the hinder half of the abdomen. Similar thoracic plates have been found in Mastodonsaurus and many other genera, while the rows of abdominal scutes are known to characterise several recently discovered forms. The reversal of the pattern found in Archegosaurus is not constant throughout the order. Several German anatomists adduce the ventral armour of the Labyrinthodonts in support of their reference of the order to the true Reptiles, and dwell upon the term 'naked Amphibia,' commonly used in German zoology to designate what English naturalists simply call Amphibia, as proof of the incompatibility of such defences with the very definition of the class. It is to be noted, however, that some few existing Amphibia possess rudimentary shields and scutes, whilst no exact parallel to the Labyrinthodont armour is to be met with in any existing animal whatsoever. A third point of importance in the organisation of Archegosaurus is the notochordal vertebral column. The axis, prefiguring what in most vertebrates becomes segmented and ossified into the centra or bodies of the vertebræ, retains even in the largest specimens of Archegosaurus its primitive or em-

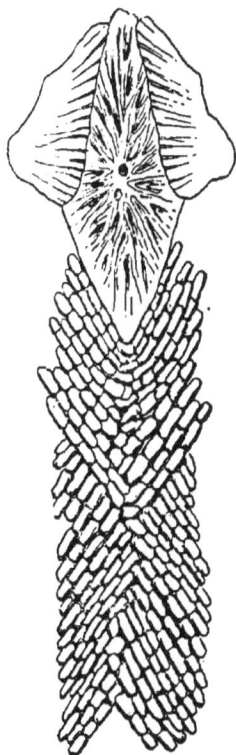

FIG. 18.   Ventral armour of Archegosaurus.

bryonic structure. There is thus no continuous spinal
column in the fossil; but the regular distances of the
bony vertebral arches testify to the presence in the
living animal of a cartilaginous rod which connected
them with each other. The occipital condyles appear to
have been similarly unossified. To prevent an unfair
comparison of these embryonic features of the Carboniferous Archegosaurus with the relatively adult structure of the Labyrinthodonts of the Trias, it should be
noticed that there are many Carboniferous Labyrinthodonts with fully ossified vertebræ.

A series of ossicles, forming as clerotic bony ring, has
been found in the orbit of Archegosaurus, and may be
expected to recur in other Labyrinthodonts. Such a
structure is unknown in existing Amphibia; it occurs in
some recent and fossil reptiles and in birds.

A genus which has since proved rather common in
the Coal Measures was first detected by Prof. Huxley,
and named by him Loxomma. The large and irregular
orbits (fig. 20) are tolerably distinctive. A close pitting,
like that seen upon the skull of a crocodile, covers all
the exposed surfaces of the cranial bones. The teeth are
remarkable for their fore and aft cutting-edges, which
are prolonged almost as far down as the imbedded base.
Similar cutting-edges are found upon the crowns of the
very young teeth of Archegosaurus and other genera. It
is not improbable that the ancestral Labyrinthodont tooth
was slender and two-edged, but that the original type is
ordinarily masked by the excessive development of the
thick-walled, cylindrical base. In Loxomma the base of
the tooth exhibits the usual labyrinthic complication,
though not to the same extent as in the large Triassic
species.

Sharp, two-edged teeth would seem to indicate that

the food of Loxomma consisted of living fish. The fish-
feeding Gavial or Gharrial of the Ganges has very similar teeth, while the Alligators and Crocodiles, animals of
more mixed diet, approach the common Labyrinthodonts
in their strong and blunt conical teeth. Very likely the
majority of the Labyrinthodonts were carrion-feeders, and
used their powerful jaws and massive teeth to crush the
bones of floating carcasses. This conjecture is rendered .
more probable by an observed difference in the articulation of the mandible. Mastodonsaurus and other large
Labyrinthodonts with crushing teeth have the lower jaw
prolonged considerably behind the condyle or hinge, as

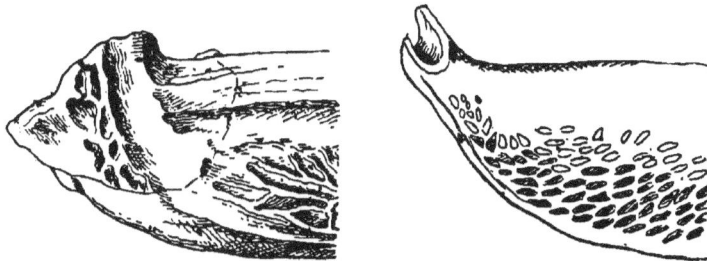

FIG. 19.   Hinder end of mandible of Mastodonsaurus (left-hand figure)
and Loxomma.

if to give additional leverage to muscles which passed
from the crown of the skull to the hinder extremity of
the jaw. Such additional leverage is well adapted to a
slow, heavy, and powerful machine. Loxomma, on the
contrary, which, if our conjecture respecting its food be
well founded, needed above all things rapidity in the
movement of its jaws, has the condyle at the very end of
the lower jaw, and loses whatever mechanical advantage
may be due to the post-articular process. It gains in
time what it loses in force.

The accompanying illustration (fig. 20) shows some
details of the skull of Loxomma,—the ' mucous grooves '

upon the face, which occur in most Labyrinthodonts, but
vary much in pattern and degree of development; the
external nostrils; the large and irregular orbits, which,
as in recent Crocodiles, no doubt lodged muscles, glands,
and other soft tissues, besides the eye-balls themselves;
and the two small rounded occipital condyles. Loxomma
has well-ossified biconcave vertebræ, and long ribs with
two articular cups. The other parts of the body are not
known for certain. In size it far surpassed Archegosaurus,
but fell nearly as far short of Mastodonsaurus, at any
rate in the dimensions of the skull. If something like
the same proportion obtained between the skull and the

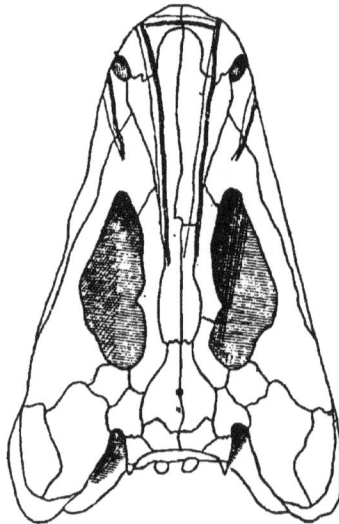

FIG. 20.   Skull of Loxomma.

total length as in the recent Gavial, Loxomma must have
been as much as fourteen feet long.

   Another Labyrinthodont of special interest is Urocordylus, first described by Prof. Huxley from remains found
in the Kilkenny coal-field. Urocordylus is distinguished
by the great length and vertical depth of its tail, which

far exceeds the rest of the body in both of these dimensions. Each of the very numerous caudal vertebræ passes
upwards and downwards into a wedge-shaped 'spine' or
process, which is flattened from side to side. The apex
of the spine is united to the vertebra, the free edge is
slightly notched, and the height is in most about twice
the extreme breadth. It will be seen at once that caudal
vertebræ of this form and to the number of seventy-five
(there are but twenty in the trunk) represent a powerful
swimming organ. Urocordylus is the most specially
aquatic of the known Labyrinthodonts. Its ventral
armour, in addition to the usual thoracic plates, consists

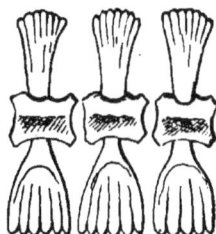

FIG. 21.   Three caudal vertebræ of Urocordylus.

of very many slender bony needles arranged in oblique,
converging rows. The skull is unfortunately imperfect
in the best British examples. Urocordylus has been
found also in North America, but is there described under
other names.

Many other Amphibia have since been found in the
Coal Measures, and at least fifteen well-investigated
species are on record. We cannot attempt to compare
and distinguish these, or to describe the discoveries made
in other parts of the world, but we may very shortly notice
a few of the remaining British genera. Anthracosaurus
resembles the large Triassic Labyrinthodonts in its
strongly sculptured cranial bones, and in the post-articular process of the mandible, as well as in the size and

internal structure of the teeth. Pteroplax, a highly interesting form as yet imperfectly known, is
apparently deficient in the lateral parts of
the skull. There is no outer bony wall to
the orbit, and no lateral row of teeth in the
upper jaw. Among recent Amphibia, Siren,
Proteus, and Menobranchus exhibit a similar
incomplete ossification of the parts between
the eye and the sides of the mouth. Batrachiderpeton resembles Pteroplax in the
deficiency of bony maxillæ. Like the recent
Siren, Axolotl, Frog, and many fishes, it
bears a mass of clustered small teeth upon
the roof of the mouth. Ophiderpeton and
Dolichosoma are believed to want both fore
and hind limbs, but the point is naturally
difficult of proof in the case of any fossil
animal, and additional examples are required to confirm or refute the supposition.
The Ophiomorpha, an order of existing Amphibia which includes Cœcilia (fig. 22) and
three other genera, are similarly destitute
of limbs.

The effect of these discoveries in pro-
moting new speculations on the one hand,
and on the other in exploding those of a
previous generation, has given them an interest beyond that which ordinarily attaches
to the labours of the palæontologist. In a
later part of this chapter we shall have occasion to discuss certain views which have

FIG. 22.   Re-
cent Cœcilia. been suggested or enforced by these and
similar revelations of extinct animals. It
will be convenient to notice here a baseless but wide-

spread notion respecting the atmosphere of the Coal
Measures, which was first effectively refuted by the
announcement that air-breathing vertebrates had undoubtedly lived during the Coal period.

Some of our readers have doubtless read with pleasure
Hugh Miller's eloquent description of the scenery of the
Coal Measures,[1] and will remember the 'scarce penetrable
phalanx of reeds,' the palm-like ferns, the pines,
the Stigmaria, which as here painted—like a gigantic star-fish, considerably exceeding forty feet in
diameter, clothed with hollow, cylindrical leaves, which
stand out like prickles of the wild rose upon the red,
fleshy, lance-like shoots—may well have seemed a 'monster of the vegetable world.' The animal life of the Coal
Measure swamp is painted with no less vividness. ' The
rank steam of decaying vegetation forms a thick blue
haze, that partially obscures the underwood ; deadly lakes
of carbonic acid gas have accumulated in the hollows ;
there is silence all around, unintrerupted save by the
sudden splash of some reptile fish that has risen to the
surface in pursuit of its prey, or when a sudden breeze
stirs the hot air, and shakes the fronds of the giant ferns
or the catkins of the reeds. The wide continent before
us is a continent devoid of animal life, save that its pools
and rivers abound in fish and mollusca, and that millions
and tens of millions of the infusory tribes swarm in the
bogs and marshes. Here and there, too, an insect of
strange form flutters among the leaves. It is more than
probable that no creature furnished with lungs of the
more perfect construction could have breathed the atmosphere of this early period, and have lived.' Hugh Miller
was here only putting into poetry what had been main-

[1] *Old Red Sandstone*, chap. xiv.

tained in argumentative prose by Adolphe Brongniart.
The supposed exuberance of primitive vegetation was
held to imply an excess of carbonic acid in the air, which
not only enabled plants of gigantic stature to exist at a
time when the earth was thinly covered with soil, but
protected their fallen trunks and leaves from rapid decomposition. The excess of carbonic acid, it was believed, would have proved fatal to air-breathing animals,
until it was condensed into the profuse vegetation of the
coal, and this explained the entire absence of such animals
from the Coal Measures.

Such was the hypothesis advanced by men of well-
deserved scientific fame, and even to this day pretty
widely current outside the narrow circle of geological
students. We may feel some wonder that an atmosphere
densely laden with carbonic acid was not seen to be in-
compatible with the existence of animals of any kind,
insects and fishes just as much as birds and beasts. This
difficulty was not perceived, but the discovery of Amphi-
bians in the Coal Measures knocked the hypothesis to
pieces. There is now no serious discussion of supposed
atmospheric differences between the Carboniferous time
and our own, and the framers of 'ideal restorations' who
exert their imagination to unite in one view all the living
things found fossil in each several formation, and to extend throughout the world such conditions as can be
shown to have obtained in any part of it, now place
Archegosaurus and its kindred prominently forward in
that landscape which was once entirely filled by Lepidodendron, Sigillaria, and the Calamites.

The two occipital condyles and the branchial arches
have been mentioned as characters which unite the
Labyrinthodonts to Amphibia, and decisively separate
them from Reptiles. A third noteworthy feature is the

presence in all Labyrinthodont skulls of a parasphenoid
ossification in the roof of the mouth (fig. 23, *par.*). Such
a bone is universal in Amphibia and in all fishes which
have an osseous skull; it never occurs as a separate bone
in the higher vertebrata, and is nearly always either rudimentary or absent. The absence of fin-rays and the
disposition of the bones of the limbs into segments comparable to those of a Reptile, Bird, or Mammal, are the
features which most clearly cut Labyrinthodonts off from
fishes. Among Amphibia no other known forms exhibit
tooth-structure like that of the Labyrinthodonts or a

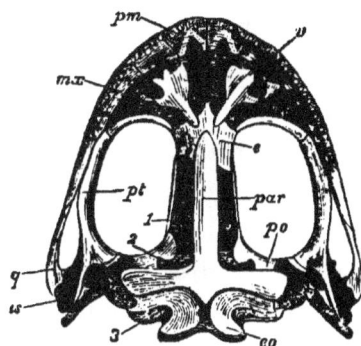

FIG. 23.   Under surface of skull of Frog, *par*, parasphenoid (after
Parker).

ventral armour, or an orbital ring like that of Archegosaurus. The zoological place of Labyrinthodonts is
satisfactorily determined by these considerations; they
form a peculiar and extinct order of the class Amphibia,
to be ranked side by side with the recent orders of
Batrachia (frogs and toads), Urodela (salamanders, newts,
&c.), and Ophiomorpha (Cœciliæ).

Some writers, and among these the eminent palæontologist Hermann von Meyer, have attached classificatory
importance to the points of likeness between the Labyrinthodonts and the Crocodiles. In the larger Laby-

K

rinthodonts, as in Crocodiles, the skull is covered with a pitted sculpture, which is probably connected with the attachment and nutrition of the tough, leathery skin. Again, the skull of a Labyrinthodont appears to be fully ossified (at all events, externally) in very young examples. No interspaces or 'fontanelles' occur in the smallest skulls of Archegosaurus, though some of these are not more than one-twentieth of the length ultimately attained. In Crocodiles the same thing is observed. Not only are the sutures of a Crocodilian skull closed at the time the egg is hatched, but the frontals and parietals, originally paired bones, are respectively united at that early period. This rapid formation of a firmly articulated skull does not preclude the further growth of every separate bone. In both Crocodiles and Labyrinthodonts the skull becomes many times larger than it was at birth, retaining all the time its accurately closed sutures, and increasing by additions to all the borders of each ossification. This mode of enlargement is compatible with great changes in the proportions of the skull. The face ordinarily grows faster than the brain-case, and the orbits may recede from the centre to the hinder third or fourth of the skull. A third point of resemblance may be discovered in the armour of the Labyrinthodont, which, though unlike that of Crocodiles in its position and arrangement, finds there its nearest parallel.

No recent Amphibian can be named which resembles the Labyrinthodonts in the points just mentioned, and it may be asked why we pass over conspicuous characters like these—characters moreover of some physiological importance—in order to dwell upon obscure details, such as the number of occipital condyles and the presence or absence of a parasphenoid bone.

A complete answer would take us far into the philoso-

phy of classification, and only the sketch of a complete
answer can be attempted here.   Let us first ask
what is the object of any classification of plants or
animals. We shall be told in reply that it is to summarise
propositions respecting structure, to frame an index in
which every species can be readily found, but above all
to illustrate the affinity of living things.   What then is
'affinity'?   Some take it to mean merely resemblance of
structure, and thus deprive the word of all special significance; others treat it as a quite unintelligible but nevertheless real bond which connects individuals of the same
species, species of the same genus, genera of the same
order, and so forth.   All who accept the common descent
of plants or animals now belonging to different species,
understand 'affinity' to mean *relationship*, and use the
word quite literally, just as they would speak of affinity
or blood-relationship among human kindred.

Such naturalists regard the subdivisions of a zoological
or botanical classification not merely as abstractions or
fictions of the human mind, still less as permanent facts of
nature, coeval with the origin of living things, but as
groups which have arisen and become definite by the
extinction of connecting links.   Good subdivisions cannot
be constructed by any process of general reasoning; they
may be discovered but not invented, they are the products
of a long chain of conditions and changes; and a mind
powerful enough to disentangle the effects of myriads of
causes acting through myriads of years might discover in
a truly natural class or order some faint reflection of the
history of the world.   Such groups exist by virtue of the
gaps occasioned by extinction; they are discovered
tentatively, and generally after many fruitless attempts;
they are recorded and identified by means of verbal definitions.   Of the characters which distinguish any natural

group the most fundamental are those which survive
many changes, and are therefore the most likely to be
inherited from some remote common ancestor. On the
other hand, such characters as immediately depend upon
some temporary peculiarity in the external conditions or
mode of life of the organism are more quickly developed
and more quickly lost. We may roughly distinguish
between ' ancestral' and ' adaptive ' characters—terms of
course relative, and admitting of degrees.

What would be ancestral characters in the individual
may be sometimes regarded as adaptive in the species,
just as what we call a long time as a fraction of the day
is a short time in comparison with the year. Similarly a
relatively permanent and fundamental specific character
may be too transitory and occasional to enter into the
definition of the class or order.

The more purely adaptive characters may prove
nothing as to affinity. Suppose it necessary to decide
whether two men are nearly related, and that all documentary evidence and testimony are wanting. Proof
might be sought in the family likeness of the two. It
would be important to show that both had red hair, or six
fingers on each hand, or that both were subject to the
gout. But it would prove nothing that both could speak
French, or play upon the violin, or that each had lost a
leg. These would be ' adaptive' characters, and the
utmost effect of many such coincidences would only tend
to show that the two men had been placed in similar
circumstances.

· In the same way the naturalist is often obliged to
estimate characters on a quite other principle than their
physiological value or outward prominence ; to neglect
or subordinate such adaptive resemblance as the whale
bears to a fish, the shrew to a mouse, the Euphorbia to a

Cactus, or the pitcher of Nepenthes to the pitcher of Sarracenia, and to lay stress upon obscure hints made known by the poring work of the anatomist, such as those which show affinity between the elephant and the manatee or rat, between the barnacle and the shrimp, and between the baobab and the mallow.

If we apply these reflections to the present case, we shall see why but little importance is attached to cranial sculpture or early ossification of the roof of the skull, when it is a question of referring the Labyrinthodonts to their class. It would be the same with the defensive bony armour, or the strengthening of the tooth by convolution of its walls. These are 'adaptive' characters, somewhat closely dependent upon mode of life, and rapidly influenced by natural selection. By virtue of similar mode of life animals of very different ancestry might come to resemble each other in these respects. But the presence of a separate parasphenoid is, so far as we can tell, of no physiological importance, and unlikely to vary with changed conditions. It therefore retains unimpaired its validity, as proof that the lines of descent of the recent Amphibia and the Labyrinthodonts have been coincident more recently than the lines of descent of the Amphibia and those higher vertebrates in which the parasphenoid is wanting or not separately developed.

The range of the Labyrinthodonts has been extended by fresh discoveries into more recent formations, and recognised examples are now recorded not only from the Coal Measures and the Trias, but from the Permian, the Rhætic bone-bed of Aust, and the Oolite of Russia. They have been discovered, too, far beyond the limits of Europe, in North America, Australia, South Africa, and India.

All Amphibia of whose habits we have any direct knowledge are land or freshwater, and this furnishes a strong presumption that the Labyrinthodonts were not marine but fluviatile. It is true that their remains have often occurred in marine limestones and conglomerates, but this may prove little or nothing as to habitat, for freshwater animals are liable to be drifted into a tidal estuary, a salt lake, or the sea.

The well-investigated Fishes of the Coal Measures belong to two groups, the Elasmobranchs (Sharks and Rays) and the Ganoids. It is not known that Teleostean fishes, such as form the great majority at the present time, are represented in the Carboniferous rocks at all.

The Sharks and Rays are eminently marine in their habits, and are so rarely found in fresh waters that their remains are quoted as almost absolute proof of the marine origin of any strata in which they occur. The known exceptions to the rule are, however, more frequent than is generally supposed. The lake of Nicaragua, which is now entirely fresh, contains, besides purely fluviatile animals, living sharks and dog-fish. There is reason to suppose that the lake has in comparatively recent times changed from salt water to fresh, and it would seem that a part of its former population has contrived to survive the change. In the Philippines a similar mixture of fluviatile species with saw-fish and dog-fish has been recorded. A species of ray inhabits the Sutlej, a shark occurs in the fresh waters of Viti Levu, one of the Fijis, and saw-fish are said to frequent the Senegal and Zambesi beyond the limit of the tide. Torpedos and sting-rays are found in the fresh waters of the Amazon, Magdalena, and other great rivers of South America.

Many of the Carboniferous Sharks and Rays are represented only by spines, such spines as the Piked Dog-fish bears in front of each dorsal fin, or the Sting-ray upon the middle of its tail. The large obliquely furrowed spines

known as Gyracanthus are often worn at the tip, and more
upon one side than the other. It is probable that they
were attached to the pectoral fins of a ground-feeding
shark, and became worn by rubbing against the bottom.
Right and left spines, each worn obliquely, but so that the
two make a symmetrical pair, have been found. The
spines, whether dorsal or pectoral, of Elasmobranchs can
be distinguished by the imbedded base, which is smooth,
tapering, and usually hollow. In Teleostean fishes on the
contrary, the base of a large dorsal or pectoral spine, such
as those of the File-fish or the South American Siluridæ,
is adapted for articulation with other bones, and occasionally exhibits a very complex interlocking mechanism.
Teeth of Elasmobranchs occur but rarely in the Coal
Measures; still more rarely are they found associated with
other parts of the same species. The shark known as
Xenacanthus, however, unites the spines formerly called
Orthacanthus with the teeth named Diplodus.

Respecting the Ganoids of the Coal Measures, it is not
easy to decide whether they were marine or fresh water.
Some may have been marine and others freshwater, or
they may have made seasonal migrations from salt water
to fresh. All the recent species are fluviatile except the
Sturgeon, which spends the winter in the sea and only
ascends rivers at the time of spawning, but there is reason
to suppose that this restriction to fresh waters is exceptional and peculiar to modern times. The Ganoids of the
secondary period were no doubt largely marine. They
occur abundantly and in excellent preservation in Liassic
and Oolitic Limestones, as well as in the Chalk, where
traces of land or freshwater organisms are very rare. The
remains associated with Ganoids in the Coal Measures do
not at once clear up the difficulty. Teeth and spines of
sharks and rays are occasionally found side by side with

Ganoid fishes; so also are bones of the eminently fresh-water Labyrinthodonts. But the clearest evidence on the point is drawn from the distribution of the species of fossil fishes throughout the Carboniferous formation. We have in England a lower series, consisting of the Mountain Limestone and associated rocks, which is altogether marine; and an upper series, inclusive of the Coal Measures, whose mode of deposition has been questioned. If the Coal Measure fishes are marine, it would be fair to expect a considerable number of species common to the upper and lower series. This does not actually occur. Ganoids are extremely rare in the Mountain Limestone, but teeth and spines of Elasmobranchs abound in particular localities. In the Coal Measures on the contrary, the Ganoids are plentiful, but remains of Elasmobranchs, excepting a few species which are almost unknown in the limestones, are scarce and restricted to certain beds of exceptional formation.

Many Ganoid fishes, both recent and fossil, are protected by an armour of bony and sometimes of enamelled scales. This feature was indeed considered universal and characteristic by Agassiz, the founder of the order. Modern zoology cannot accept such definition as adequate. There are true Ganoids without bony scales, and there are armour-plated fishes which are not Ganoids. Perhaps the order cannot be strictly defined, seeing that a majority of the species attributed to it are extinct and but imperfectly known. Study of the recent species has, however, brought to light characters which are distinctive of them, and perhaps of many of the fossils also. All existing Ganoids have a swim-bladder and a spiral valve in the intestine. The ventral fins, if present, are abdominal, that is, set far back. The optic nerves, instead of simply crossing, as in our common fishes, form a chiasma (χ) with slight inter-

mingling of their fibres; this occurs also in the Elasmobranchs and in the higher vertebrates. Affinity between
the Ganoids and Elasmobranchs is also shown by the
occurrence in both of sets of valves in the contractile tube
(*bulbus arteriosus* or cardiac aorta) which leads from the
ventricle of the heart, but in Ganoids these valves are
usually more numerous and arranged in a greater number
of rows.   Free gills and a gill-cover distinguish the
Ganoids from the Elasmobranchs.

We can attempt no description or even enumeration

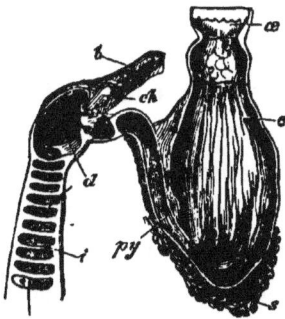

FIG. 24. Section of Stomach and
Intestine of a Shark (*Squalus
maximus*); *i*, spiral valve in intestine.

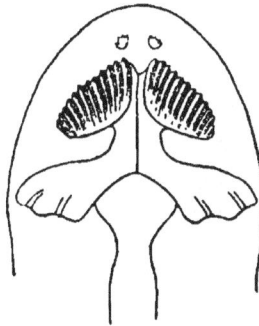

FIG. 25.   Under surface of skull
of Ctenodus.

of all the genera and species of Ganoids found fossil in the
Coal Measures.   A very few may be selected for particular
notice, and of these we shall first consider Ctenodus and
its allies.

Ctenodus was first figured and described by Agassiz in
his great work on Fossil Fishes, from a tooth found near
Leeds and still preserved in the Leeds Museum.   This
fossil, of which some notion may be got from fig. 25, is
semi-oval and marked by about eleven transverse ridges.
Each ridge is saw-like, and rises into points towards the
outer or straight side.   Two such plates are set in the roof

of the mouth, and separated by a slight interval. It has
since been made out that a similar pair of crushing teeth
was attached to the lower jaw, while Mr. Atthey has
found some small serrate teeth, which are very likely the
' vomerine' teeth of the same fish, and were originally set
in advance of the two palatal plates. The skull, scales,
and other parts of Ctenodus are more or less perfectly
known, but the dentition is the most characteristic
feature; it consists, as we have seen, of two opposed pairs
of dental plates, each crossed by transverse ridges, and of
a pair of small pointed vomerine teeth, the plates being
adapted to crush and grind, the vomerine teeth to pierce
and cut.

Long before Ctenodus was known, a small fish from
the Old Red Sandstone had been discovered and named
Dipterus. Its structure was first adequately revealed by
Hugh Miller in his ' Footprints of the Creator,' and subsequent investigation has not added much to his description, though it has given significance to various details,
which could not be seen in their true light thirty years
ago. Dipterus has a pair of triangular plates in the roof
of the mouth, and these bear radiating ridges, but the
ridges are broken up into points by intersecting grooves.
A similar pair of plates is attached to the lower jaw.
Thus the grinding apparatus is essentially the same as in
Ctenodus, and quite recently the existence of a pair of
vomerine teeth has been established in Dipterus also.
Such teeth have not, it is true, been actually seen, but
Dr. Günther has figured a Dipterus skull which shows
quite unmistakably the two small sockets, and these
exactly where the vomerine teeth ought to be. Dipterus
is further distinguished by the form of the paired fins,
which are long and slender, covered along the centre with
a thin, pointed strip of scales and fringed by fin-rays.

Agassiz has described in his book on Fossil Fishes some large and peculiar teeth found in the Trias of Germany and also in a conglomerate or 'bone-bed' of about the same geological age, which occurs at Aust Cliff on the Severn. These teeth were triangular or semi-oval in outline, with blunt transverse ridges and a wavy margin. They were at first believed to belong to Sharks, and were named Ceratodus or 'horn-tooth' on account of their projecting cusps. Agassiz conjectured very acutely (for none of these teeth were united, or retained more than a

FIG. 26.   Under surface of skull of Dipterus.

trace of the attached bone) that they had been set in opposed pairs in the mouth, and that the ridged or wavy margin had been outermost. These suppositions have been amply confirmed. In 1870 Mr. Gerard Krefft found a quite similar fish living in the fresh waters of Queensland. Many specimens have now been examined by zoologists, and it is placed beyond doubt that the genus Ceratodus still survives. The crushing plates are curiously like the fossil teeth of the Trias, and set as Agassiz had supposed, one pair in the roof of the mouth, another in the lower jaw. The recent Ceratodus moreover exhibits a pair of cutting vomerine teeth, in front of the palatal plates, just as in Dipterus and probably also in Ctenodus.

There are thus three distinct forms, belonging to three or
more geological periods, which resemble each other in a
very exceptional arrangement of teeth. The resemblance
does not end here. Ceratodus has paired fins outwardly
similar to those of Dipterus. There is in each the long

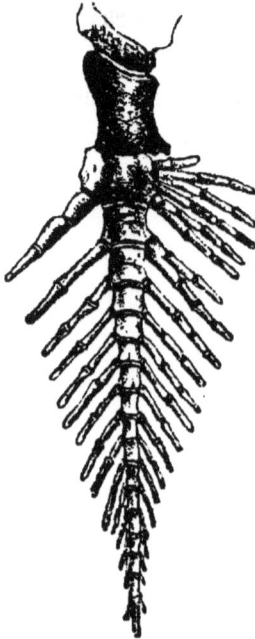

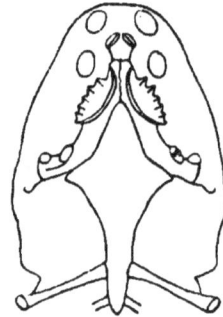

Fig. 27.   Cartilaginous framework   Fig. 28.   Under surface of skull of
       of pectoral fin of Ceratodus.   Ceratodus.

patch of scales fringed by fin-rays, a structure almost as
rare as the dentition itself. Beneath this outward covering the scalpel has revealed in Ceratodus a cartilaginous
framework of still greater novelty and interest. A long
jointed rod runs along the centre of the limb, and gives
off slender, tapering, and jointed appendages on either
side, one (usually) to each segment of the central axis.
But for the exceptions to be noticed immediately, no other
existing fishes exhibit a like structure; the external re-

semblance, however, of the paired fins of Dipterus and some few other fossil fishes renders it highly probable that they agreed with Ceratodus in the internal frame-work of the limbs. Again, the recent Australian fish was found to have a peculiar structure of the nostrils. The olfactory organ in most fishes is a sac, opening externally, but having no communication with the mouth. In Ceratodus, as in Amphibia and all the higher vertebrates, the nostril is not a mere sac, but a tube opening into the palate by an oval aperture on each side of the mouth. If we now turn to Hugh Miller's figure of the roof of the mouth in Dipterus,[1] we see a similar pair of oval apertures in the same position, and this establishes another interest-ing point of resemblance between Ceratodus and Dipterus. Direct proof of internal nasal passages in Ctenodus is not as yet accessible.

Some forty years ago zoologists were much interested, and at the same time not a little perplexed, by the de-scription of a new animal sent to Europe from South America, and named Lepidosiren. A fish in outward appearance, it possessed at the same time paired nostrils communicating with the mouth, and cellular lungs with an air-duct and glottis. The pectoral and ventral fins were represented by slender filaments. Two years later a nearly allied animal from Senegambia was examined by Prof. Owen. This is sometimes known as Protopterus, but other naturalists treat it as a second species of Lepi-dosiren. The definitions then extant were incompetent to settle whether these animals were Fishes or Amphibia, and a long controversy ensued, during the course of which every point of their structure was eagerly scrutinised. Naturalists are now agreed that the view propounded by Prof. Owen is in the main correct, that Lepidosiren is a

[1] *Footprints of the Creator*, fig. 20.

fish, and that the definition of a fish must be extended to
include vertebrate animals with functional lungs and
nostrils communicating with the mouth. In Lepidosiren
and Protopterus these structural peculiarities are connected with the external conditions of life. They inhabit
intermittent streams or pools, and as the waters retreat
the fishes are liable to be left dry upon the bank. When
this happens they bury themselves in the mud, leaving a
small aperture for air-communication. The mud dries
and cracks in the sun, and the fishes become enveloped in
earthy cases lined by a thick mucous secretion. In this
state they remain till the return of the rainy season
enables them to escape and swim about again. During
the torpid state respiration is carried on by means of the
lungs, which temporarily aerate the blood; at other times
the gills oxygenate the blood in the usual way. The
habits of Ceratodus are less perfectly known than those of
Protopterus (the African species), but it is said to leave
the water at times, and its lungs are well developed and
adapted for aerial respiration. In all three cases it is to
be noticed that the lung is merely a special modification
of the swim-bladder. We have thus fresh proof of the
homology of the swim-bladder of the fish with the lung of
the air-breather, a doctrine long ago maintained by
Harvey and Hunter, and now clear beyond dispute.

Two points in the structure of Lepidosiren and Protopterus more immediately concern us here. The first is
that the pectoral and ventral filaments are supported by
a slender cartilaginous rod, which finds its nearest parallel in the fin of Ceratodus. Even before the discovery
of the recent Ceratodus, Prof. Huxley was sagacious
enough to remark that ' Lepidosiren is, in fact, the only
living fish whose pectoral and ventral fins have a structure
analogous to that of the acutely lobate paired fins of

Holoptychius, of Dipterus, or of Phaneropleuron, though
the surface-scales are still less developed in the modern
than in the ancient fish.' The dentition, in a more obvious
manner, connects all the forms now under consideration.
Lepidosiren and Protopterus have, like Dipterus, Ctenodus,
and Ceratodus, opposed pairs of crushing plates, but the
ridges are more prominent and sharper; there is also a
pair of conical vomerine teeth placed as in Ceratodus or
Dipterus.

It will be seen that there are good grounds for placing
the three recent genera (Lepidosiren, Protopterus, and
Ceratodus) and the two fossil genera (Ctenodus and
Dipterus) near together. One character to which importance is attached, separates Dipterus from the rest. It
has what is usually called a heterocercal or unequally
lobed tail, the vertebral column being prolonged into the
upper lobe. The tail of Ctenodus is not known. In the
three recent genera a symmetrical tapering tail occurs, of
the kind sometimes called diphycercal. Where are the
five associated genera to be placed? Many zoologists
have deemed the possession of functional lungs a character of sufficient weight to justify the creation of a separate
order (Dipnoi) for these fishes, but this is plainly an
'adaptive character,' and though important, is not important enough to serve as the foundation of an order or any
major group. We shall perhaps hereafter discover fishes
of quite different affinity which have a swim-bladder
adapted to aerial respiration. Any test of natural history
characters should of course be applied impartially, and it
may be suggested that we have not allowed for adaptation
in the case of the dentition—that the dentition is as adaptive as the lung, that it conforms to a peculiar diet, and
should therefore be discarded as a chief character. These
considerations have some real weight, but it is to be

observed that the adaptation of the teeth is highly special. An elaborate mechanism is closely repeated in all the genera, while the non-essential details (such as the attachment of the teeth to particular bones) are repeated as well as the essential ones. It would be most improbable that a quite similar seizing and crushing apparatus should ever arise in another and independent group of animals, nor has any example actually occurred of a fairly comparable structure in a fish of different affinities. The nearest approach to such a case is perhaps met with in the fish known as Chimæra; but here all the details, and especially the mode of insertion of the teeth, are different. Returning now to the question of the place of these five genera, we have to notice a series of important anatomical features which all the recent species exhibit. There is a swim-bladder (converted into a lung), a spiral valve in the intestine, abdominal ventral fins, an optic chiasma, valves in the *bulbus arteriosus*, free gills, and a gill-cover. These are the very characters which define the Ganoid fishes (see p. 136), and if we may for the reasons given treat the temporary respiratory difference as of subsidiary value, there is no reason why all five should not rank as Ganoids. Even in respect of the lung the separation is not very emphatic, for the swim-bladder is cellular and double in the Ganoid Polypterus, and communicates with the pharynx by a passage which nearly resembles the air-duct of Lepidosiren or Ceratodus. Other Ganoids share more or less in the same structure. These fishes, and some few other fossil genera of which we can give no account here, may be regarded therefore as a peculiar group of Ganoids. The paired fins, with the long, narrow, central patch of scales fringed by fin-rays, and hence called ' acute-lobate,' form perhaps their most constant feature, and serve to connect with them certain fossils whose internal organisation is quite unknown. Gegenbaur has proposed the name of *archipterygium* for the primitive and fundamental type of limb of which Ceratodus furnishes the completest example yet known. We may conveniently use this term to designate the group which at present goes without name; and rank the five genera, Lepidosiren, Protopterus, Ceratodus, Ctenodus, and Dipterus, with other 'acute-lobate' forms, under the common name of Archipterygian Ganoids.

The group ranges in Europe from the Old Red Sandstone to the Trias, and has even been traced into the Oolite. Here it disappears, and might, so far as the evidence of European rocks is concerned, be considered extinct. But elsewhere kindred forms survive, and perpetuate a singular type which dates almost as far back as the oldest known vertebrates. There is perhaps no equally striking case of persistence on record, for the Archipterygian Ganoids are not low in their class, like the Foraminifera and other often-cited examples, whose indefinite duration may receive explanation from their slight degree at once of differentiation and integration, and from their comparatively feeble sensibility to physical changes.

Not less remarkable is the range in space of the Archipterygian Ganoids. They have now been found fossil in many parts of Europe, in India, in South Africa, and in North America, while they survive in tropical Africa, South America, and Australia. To connect these localities in their order of succession and to trace the migrations of the group will some day prove to be an instructive and not altogether hopeless problem, but as yet we want both the facts and the methods necessary to a satisfactory solution.

One of the commonest and most conspicuous of the

L

Coal Measure Ganoids is Megalichthys. Its large, squarish scales, brilliant with enamel and closely set with microscopic pores, are often scattered through the débris of a shale-heap. The head is defended by larger plates of similar texture. Between the halves of the lower jaw are two principal (besides several lateral and one median) 'jugular plates,' which replace the branchiostegal rays of Teleostean fishes and of many Ganoids (see p. 149). The paired fins are covered at their bases by patches of small scales, but the patch or 'lobe' of scales ends in a rounded edge, instead of tapering nearly to the tip of the fin, as in the Archipterygian Ganoids. The two types of fin-structure are sometimes distinguished as 'obtuse-lobate' and 'acute-lobate.' When the recent Polypterus, which is obtuse-lobate, is compared with the acute-lobate Ceratodus, the internal framework of the pectoral fin is found to be altered almost beyond recognition. The long, tapering, jointed axis seems to be reduced to a single cartilage, flanked on each side by a slender bone, and these three basal pieces carry a transverse series of radials, which are embraced by the fin-rays. The archipterygian type is not plainly visible in such a limb as this. Polypterus, and doubtless the other obtuse-lobate Ganoids also, are indeed intermediate, as to the internal structure of their paired fins, between the Elasmobranchs and the Teleostei. Experience has shown that the structural difference, of which the obtuse and acute lobes are the outward indications, possesses a considerable systematic value, and in our present uncertainty as to the natural subdivisions of the Ganoid order, we are glad to adopt this distinction as likely to prove one element of a permanent arrangement. Megalichthys has an imperfectly ossified vertebral column; the circumference only of each centrum is ossified, so that in the fossil state the vertebræ form rings of perhaps an inch diameter. The teeth are conical, pointed, and very unequal; towards their bases a convolution of the tooth-wall, analogous to that of the Labyrinthodonts but less complicated, may be made out. Patches of small clustered teeth occur on the vomers in the roof of the mouth, as in some Labyrinthodonts. (See Batrachiderpeton, p. 126.) Megalichthys attained a length of several feet; the skull by itself has been known to measure twelve inches by nine.

Rhizodus, an allied genus, may have been still larger. The head is covered with rough, granulated scutes, and the body with large rounded scales. The teeth are two-edged, not unlike those of Loxomma (p. 122), and very unequal in size. Some of the largest are occasionally from two to three inches long, and nearly an inch in diameter at the base.

Scales of the Ganoid Cœlacanthus are by no means infrequent in coal shale, but specimens exhibiting any considerable part of the fish are rare. The body is small, rarely so much as a foot long, and covered with thin, rounded scales upon which a surface-ornament, made up of converging ridges, may be distinguished. There is a large symmetrical (diphycercal) tail, and the vertebral column passes through its centre, ending in a slender tip, which bears a few small fin-rays as a kind of supplementary tail. The vertebræ are unossified (notochordal) as in many palæozoic fishes, and the ribs are either absent or inconspicuous. The obtuse-lobate paired fins resemble those of Megalichthys or the recent Ganoid Polypterus, and jugular plates are present as in them. Teeth of two sizes, the smaller in close patches, have been found. The chief point of structural interest is the air-bladder, which is altogether ossified—a feature unexampled beyond the

limits of the Cœlacanth family, and without adequate
physiological explanation.

In Great Britain and Western Europe few genera of
animals or plants are found common to both palæozoic and
secondary rocks, and a very decided palæontological line
is therefore drawn at this point. We may well suppose
that this is owing to the exceptional circumstances
under which the British Triassic and Permian rocks were
formed—circumstances highly unfavourable to the existence of animal life. But in the Austrian Alps where
marine beds of Triassic age occur, and where an abundant
fauna of that period is preserved, palæozoic and secondary
genera overlap, and other mixed faunas may be discovered as distant countries become better known. Our
defective knowledge, which is partly due to the rarity
of fossils in the Bunter and Keuper Sandstones, the
lowest members of the secondary rocks of Western Europe,
may render the break more apparent in the geological
chart than in the actual history of the time. A very few
types bridge over this gap in the British fossiliferous
strata, and connect the life of the palæozoic with that of
the secondary rocks. Of these the Labyrinthodonts
furnish one prominent example, as Prof. Huxley has
observed, and the Cœlacanths another. The Cœlacanth
family persists from the Carboniferous through the
Permian to the Triassic period; it is represented by one
genus in the Oolite, and by two in the Chalk, and the
latest survivors retain all the features which define the
group at its first appearance. The Archipterygian Ganoids
described above exhibit a succession even more prolonged,
and extending from the palæozoic to the modern period.

Another important section of the Coal Measure
Ganoids includes the genera Palæoniscus [1] and Platysomus.

[1] Palæoniscus has been much split up of late, and some writers now

The first is of slender form ; the second flat, and very deep
in the side. In both the scales are enamelled, rhombic,
and strongly articulated, usually by spine and socket, to
those above, and by keel and gutter to those in the same
row. The tail is heterocercal (p. 143), and there are no
scales upon the surface of the fins. The gape of the
mouth is very extensive. No jugular plates are present,

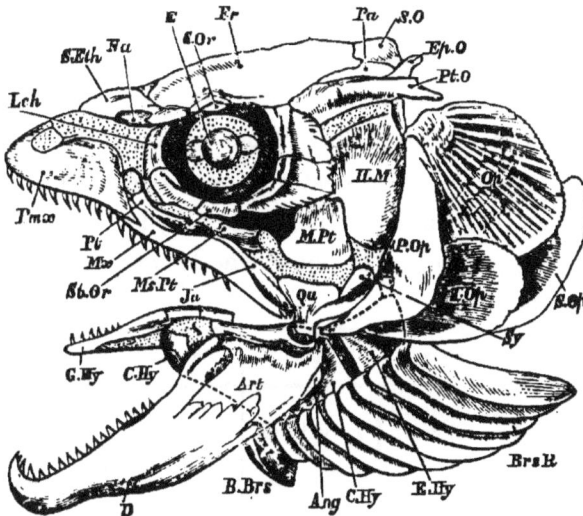

FIG. 29.   Side view of Skull of Salmon. *Brs. R.* Branchiostegal Rays
(After Parker).

but the gill-slit is bordered, as in most fishes where it
occurs, by a flexible valve of many overlapping branchiostegal rays, which closes against the clavicular arch, like
a door upon its sill. The vertebral column is notochordal
(p. 121), and there appear to be no ribs. Each fin bears
upon its fore edge a row of pointed overlapping scales
(fulcra). The teeth are slender, conical, and usually
minute, larger ones standing up at short and regular

confine the name to a few species which occur only in Permian rocks,
This is done, however, on the ground of minute differences, whose systematic
value is very doubtful.

150 COAL. CHAP. IV.

distances. Nearly every word of this description also applies to Lepidosteus, a recent North American Ganoid, which differs mainly in the complete ossification of its vertebræ, and the Palæoniscidæ are hence pretty generally regarded as forerunners of the Lepidostean line,[1] which in secondary times gave rise to many branches now extinct. The various lines of zoological succession which have here been traced among the Ganoids may now be set down in a tabular form, but it cannot be too strongly insisted that the classification of the order is still most imperfectly made out, and that as yet only tentative arrangements are possible. Some Carboniferous genera, and especially the interesting genus Acanthodes, are omitted as too imperfectly understood; others as of smaller importance. Groups wholly non-Carboniferous are also left out.[2]

	Archipterygian	Polypterine	Cœlacanth	Lepidostean
Recent . .	Protopterus Lepidosiren Ceratodus	Polypterus		Lepidosteus
Secondary .	Ceratodus		Holophagus Macropoma	Lepidotus Eugnathus Aspidorhynchus
Carboniferous	Ctenodus	Megalichthys Rhizodus Rhizodopsis	Cœlacanthus	Palæoniscus Platysomus
Devonian .	Dipterus Holoptychius Glyptolepis	Osteolepis Diplopterus		Cheirolepis

[1] Dr. Traquair, who has particularly studied the Palæoniscidæ, considers them allied most nearly to Polyodon, but an analysis of the sub-orders which he adopts seems to show that in uniting Palæoniscus and its allies to Polyodon and the Sturgeons, he breaks down all the important features which separate these latter from Lepidosteus.

[2] A thick line marks the probable limits of Prof. Huxley's sub-order *Crossopterygii*, defined chiefly by the obtuse- or acute-lobate paired fins, and the jugular plates. The chief objection to this group is that it cuts in two the Archipterygian line, which, though in some points imperfectly made out, promises to prove a thoroughly well-connected series.

The living Ganoids include species from the rivers and lakes of Europe, Africa, China, North and South America, and Australia. Of these the Sturgeons periodically visit the sea, at least in the Old World, though this is not always the case in North America; the rest are purely fluviatile. There is much uncertainty as to whether the many described forms of Sturgeon are best treated as species or varieties. Dr. Günther counts them as twenty-five, and this makes the total of living species thirty-seven.

From these particulars alone it would almost be possible to reconstruct the general history of the Ganoid order. Anciently marine as well as fluviatile, they are now almost absolutely restricted to fresh waters. Their wide distribution shows their great development in past ages; their scattered habitat, as well as the richness in types of an order numerically so scanty, is proof that they are but the relics of a race which once occupied every river and sea. More favoured forms, whose predominance it would be hard to explain, now hold much of their ancient territory. In the open sea the Ganoids can no longer hold their own, but a few species, isolated by extinction both in their organisation and their range, linger on in distant river-basins, protected from destructive competitors by the mountains and seas which now enclose them.

Of the invertebrate fossils of the Coal Measures the Mollusca, while less interesting structurally than some other groups, for they exhibit few important deviations from types otherwise known, are believed to afford the most trustworthy information respecting the circumstances of deposition. They are numerous and well-preserved in most geological formations, while throughout the whole sequence of fossiliferous rocks certain forms are found to

be restricted to marine deposits, others to fresh water,
and others again to dry land. Thus the Cephalopoda,
Pteropoda, and Brachiopoda are exclusively marine. Of
the Gasteropoda some families are marine, others fluviatile, while the Pulmonate order is nearly altogether terrestrial. The Lamellibranchiata are all aquatic, and mostly
marine, but families and genera always fluviatile are
known. Evidence drawn from fossil Mollusca is regarded
as decisive when it can be fairly applied to questions of
deposition, and in the many discussions as to the origin
of the Coal Measures the Mollusca have been eagerly
appealed to. We shall briefly summarise the Mollusca of
the Coal Measures, not merely for the sake of the light which
they may throw upon the circumstances of deposition, but
as part of a general census of Carboniferous fossils.

The Cephalopoda of the Coal Measures include the
genera Goniatites, Nautilus, Discites, and Orthoceras.
All have a chambered shell traversed by a siphuncle, and
are thus identified as members of the tetrabranchiate
division of the class. Among the Brachiopoda species of
Terebratula, Spirifera, Productus, Chonetes, Discina, and
Lingula have been found. The Gasteropoda of the European Coal Measures are all marine, but two species of the
air-breathing genus Pupa have occurred in Nova Scotia
and Illinois. Of the Lamellibranchiata all the well-understood genera (Aviculopecten, Ctenodonta, Posidonomya,
&c.,) are marine. The so-called Unios (Anthracosia, &c.,)
bear many names, but their distinctive characters are as yet
ill-traced, and little use can be made of the recorded
notices of their occurrence. This is the more unfortunate
that their distribution in the Coal Measures is peculiar.
Of Pteropods Conularia and, perhaps, Porcellia have been
found. All the genera of marine Mollusca here mentioned
occur also in the Carboniferous Limestone, while a large

and increasing number of species are common to different parts of the Carboniferous series, reappearing wherever there is a return of similar conditions. The incompleteness of our information is sufficient to explain why certain species are recorded only from the Limestones, and a few others only from the Coal Measures.

The list just quoted contains many marine forms, one or more doubtful species, which are probably fluviatile, and but one terrestrial, air-breathing genus (Pupa). We should naturally infer that the Coal Measures were chiefly marine, with perhaps a trifling mixture of land and fresh-water deposits. No conclusion could be more erroneous : and this may illustrate the necessity of studying the sources of our statistics no less carefully than the statistics themselves. If we examine the localities which have furnished this list of Mollusca, it will be found that the marine species, which so greatly preponderate, come from particular bands, of small number, of very trifling thickness, and, in all but one case, of very little constancy. The rest of the English Coal Measures, including a vast series rich in plants and fish-remains, rarely contains any Mollusca except Anthracosia, a doubtful genus, which may include several distinct forms, and whose habitat, whether marine, fluviatile, or terrestrial, is not certainly known. The list thus analysed directly proves nothing more than that certain bands, of peculiar mineral character and quite inconsiderable in number and thickness, are of marine origin.[1]

It may fairly be argued, indeed, that bands crowded with species which are elsewhere absent from the Coal

[1] An exactly parallel case is furnished by the Lower Oolite rocks of the Yorkshire coast. Here the mollusca are all marine, but nevertheless the series as a whole is pretty plainly of estuarine origin. As in the Coal Measures, the marine fauna is confined to a few thin bands, which differ from the rest both in mineral character and in fossil contents.

Measures are likely to be exceptional in their origin. If all the species found in such bands are marine, we may find in this very fact grounds for supposing that the mass of the Coal Measures is not marine. When any assemblage of species keeps totally apart from others, there is *primâ facie* reason to suspect that it differs from the rest in some circumstance which regulates distribution.

To avoid returning to this question, let us go on to consider what light can be thrown upon the origin of the Coal Measures by other classes of fossils. Setting aside the marine bands as exceptional, we shall tabulate the main fossil groups of the remainder of the Coal Measures, marking the type of distribution of each group in other geological periods and at the present time. Where a group is partly marine and partly fluviatile or terrestrial, we shall note the case as ambiguous. This done, we get the following table:

Fossils of Coal Measures, exclusive of Marine Bands.

Amphibia (Labyrinthodonts).	All fluviatile or terrestrial.
Ganoids.	Ambiguous.
Elasmobranchs.	Nearly all marine.
Gasteropoda (none British? Pupa in N. America).	Terrestrial.
Lamellibranchiata.	Imperfectly identified.
Crustacea.	Ambiguous.
Insecta.	Terrestrial.
Arachnida.	Terrestrial.
Myriapoda.	Terrestrial.
Annelida.	Ambiguous.
Plants.	Terrestrial.

The result is strongly in favour of the fluviatile or terrestrial origin of the bulk of the Coal Measures, and in adopting this conclusion we have no difficulty to face except that implied in the supposition that the Elasmobranchs were more largely fluviatile than in other periods. A more detailed knowledge of the distribution of the fossils in particular horizons of the Coal Measures would perhaps still further reduce this difficulty, by showing that all the species of Elasmobranchs are not mixed indiscriminately with terrestrial and fluviatile forms.

An attempt has lately been made [1] to apply the fossils and especially the Mollusca, of the Carboniferous Rocks to a geological classification. We are not here concerned with purely stratigraphical questions; but the stratigraphical bearing of fossils is one of the most important aspects of palæontology, and it may be worth while to indicate some preliminary considerations necessary to a right use of the evidence which palæontology has to offer.

Fossils may throw light upon the conditions of deposition, determine certain strata as marine, fluviatile, or terrestrial, as tropical, temperate or arctic, and in various other ways contribute details to that chapter of events of which stratigraphical geology should be the reflection. When fossils are thus used for geological classification, the events which they signify ought to be so conspicuous and wide-spread as to constitute epochs. Trifling and temporary physical changes are too frequent and too local for stratigraphical division. Where, as in the present case, emphasis is laid upon the difference in origin of various parts of the Carboniferous series, one set of beds being marine, others possibly estuarine, others again terrestrial or fluviatile, we must be clear that the oscillations of level thus denoted were prolonged, and of continental extent. If it can be shown that a tract of land has gone slowly and steadily down for thousands of feet, like Western Europe in early Mesozoic times, we feel sure that contiguous areas must have shared more or less in the movement. It may not have been a world-wide

---

[1] Prof. Edward Hull on The Upper Limit of the Essentially Marine Beds of the Carboniferous Group of the British Isles and adjoining Continental Districts; with Suggestions for a fresh Classification of the Carboniferous Series.—*Q. J. Geol. Soc.*, xxxiii. p. 613 (1877).

event—few geological changes are—but it must have extended its influence over a large part at least of the area
which our stratigraphical classification is constructed to
cover; it is an event at once memorable, far-reaching, and
distinct. If the limestones and marine bands of the
Carboniferous rocks are to enjoy the same consideration,
they too must be shown to have a comparable magnitude,
importance, and constancy. Enough has been said in earlier chapters of this book to put any such claim out of
question. The marine formations of the Carboniferous
system are partial and variable, both as to number and
position, and any classification resting upon them can
have only an extremely local value.

Fossils may be used as mere arbitrary or unintelligible
symbols, and nevertheless by the constancy of their succession they may have a classificatory value, like the alpha-
betical arrangement of an index. In this case still more
stress must be laid upon the regularity of their occurrence; unless the marks always return in the same order,
they cannot be made to characterise systems of any kind.
A survey of the facts recorded concerning the vertical
succession of Carboniferous fossils even in the small area
of the British Isles obliges us to conclude that there is
no such constant recurrence or alternation as would bear
the weight of a classification.

Yet again, changes in fossils may indicate life-epochs,
and hence may serve to establish relations of time. The
presence or absence of particular fossils in particular parts
of the Carboniferous rocks may be cited as evidence of a
change of life upon the earth. But to authenticate such
permanent changes, it is necessary that similarly situated
assemblages of plants or animals should be compared.
We must compare marine fossils with marine fossils, land
plants with land plants, mollusca with mollusca, and so

on, neglecting such differences as might be attributed to outward physical conditions.   This consideration has at times been overlooked, and the difference between the fossils of an ordinary coal-shale and of a marine band has been regarded as the token of a wide, if not universal, change.   There has been talk of ' a palæontological break ' between the Gannister and Middle Coals of the north of England; it is much such a break as that between the animal population of a field in Middlesex and the beach at Brighton.   True zoological zones may exist in the Carboniferous rocks, but they have yet to be traced ; nor will it be easy, after discounting the influence of conditions of deposition, of differences in preservation and of varying facilities of discovery, to collect a residuum of significant and unimpeachable characteristics.

Insects are naturally rare in the Coal Measures, as in all rocks deposited from water, but several different genera of Orthoptera and Neuroptera have been recognised. These are in most cases of curiously modern aspect, and agree in various minute details, such as the arrangement of the veins upon the wings, with recent families.   Cockroaches, termites or white ants, locusts, leaf-insects, and May-flies have been identified.   The existence of Coleoptera and Lepidoptera in the Coal Measures has been affirmed, but on slender grounds.

Scorpions and spiders are also known.   A nodule of clay-ironstone from Dudley has yielded a small 'false-scorpion,' with a ringed abdomen united to the thorax by a broad base.   A second genus of the same sub-order has occurred both in England and in North America.

Myriapods have been found in Nova Scotia, and quite recently in England and Scotland.   In Nova Scotia Dr. Dawson found the myriapod Xylobius, together with small Amphibia and the land-shell *Pupa vetusta*, crowded

together in a coaly mass which occupied the cavity of an
erect Sigillaria.

Among the Crustacea the king-crabs are represented
by Bellinurus and Prestwichia. Anthrapalæmon is not
unlike a cray-fish, but the eyes and some other features
necessary for a zoological determination, have not as yet
been seen. The occurrence of Trilobites in the Coal
Measures has been often affirmed, but conclusive evidence
is still wanting. This ancient group is scantily represented in the Mountain Limestone, and does not extend
beyond the Carboniferous period.

Comparative Table of Recent and Carboniferous Animals.

Recent.	Carboniferous.
VERTEBRATA.	
Mammalia.	
Aves.	
Reptilia.	
Amphibia.	Labyrinthodonts. Urodela ?
Pisces.	
Teleostei.	
Elasmobranchii.	Many.
Ganoidei.	Many.
* Marsipobranchii.	
* Pharyngognathi.	
MOLLUSCA.	
Cephalopoda.	Many.
Gasteropoda.	Many.
Pteropoda.	Conularia, Porcellia.
Lamellibranchiata.	Many.
Brachiopoda.	Many.
* Tunicata.	
Polyzoa.	Many.
ARTHROPODA.	
Insecta.	
Orthoptera.	Locusts, Cockroaches, Leaf-insects, White Ants.
Neuroptera.	May-flies, &c.
Coleoptera.	?
Hymenoptera.	
Lepidoptera.	

Comparative Table of Recent and Carboniferous Animals.

Recent.	Carboniferous.
Diptera.	
Hemiptera.	
Arachnida.	Scorpions, False Scorpions, Spiders.
Myriapoda.	Centipedes.
Crustacea.	
Decapoda.	Anthrapalæmon, Palæocrangon.
Stomapoda.	Pygocephalus.
Isopoda.	
Trilobita.	Brachymetopus, Phillipsia, Griffith-ides.
* Amphipoda.	
Xiphosura.	Bellinurus, Prestwichia.
Eurypterida.	Eurypterus ?
Phyllopoda.	Ceratiocaris, Dithyocaris, Leia, Es-theria, Cyclus.
Ostracoda.	Many.
* Copepoda.	
* Rhizocephala.	
Cirripedia.	
VERMES.	
Annelida.	Spirorbis, &c.
* Gephyrea.	
* Chætognatha.	
* Nematelminthes.	
* Platyelminthes.	
ECHINODERMATA.	
Echinidea.	Palæocidaris, Palæchinus, &c.
* Holothuridea.	
Asteridea.	Protaster.
Ophiuridea.	
Crinoidea.	Many.
CŒLENTERATA.	
Actinozoa.	Many Corals.
Hydrozoa.	Palæocoryne, Favosites, Chætetes, &c.
Spongida.	?
PROTOZOA.	
* Infusoria.	
Rhizopoda.	Many Foraminifera.
[Monera ?]	

₊ The fossils of the Mountain Limestone are included. * means
that the group is incapable of preservation in a fossil state.

By including in the table the marine species of the Mountain Limestone we gain a somewhat less imperfect view of the life of the Carboniferous period, which, comprising as it does marine, fluviatile, and terrestrial forms, is the first tolerably comprehensive fauna in geological history. We must be careful not to presume that our list is even an approximately complete representation of the actual Carboniferous fauna. So great are the difficulties of preservation in a fossil state that the known fossils of any geological age are only a selection, and by no means an impartial selection, of the species which then lived. This is no doubt well known, but it is constantly overlooked in practice. Brongniart, when he desired to establish the existence of a carbonic acid age, Haeckel, when he attempted to engrave the pedigree of the animal kingdom,[1] practically assumed that our knowledge enabled us to say what classes of animal had and what had not come into existence in the Carboniferous or any other period. That the assumption is dangerous and unsound is shown by a century of mistakes ; the history of palæontological speculation is strewn with the wrecks of hypotheses which have gone to pieces on this particular rock. The maxim *de non apparentibus et non existentibus eadem est ratio* may be sound in law, but it is vicious in geology.

In particular we must notice that the absence of remains of the higher vertebrates (reptiles, birds, and mammals) is no proof that they did not exist in Carboniferous times. For various reasons birds and mammals especially are very rare as fossils, except in superficial and little altered formations of terrestrial or fluviatile origin. The fact in evidence, viz. that they have not been found in the Coal Measures, warrants no conclusion whatever.

[1] *Schöpfungsgeschichte*, Taf. v, xii.

Here again we may learn wisdom from past mistakes. Fifty years ago it was sometimes maintained that quadrupeds not only did not occur in rocks earlier than the tertiary period, but that they could not have lived at the time of the formation of the secondary rocks. Fresh discoveries however have gradually demonstrated the existence of quadrupeds in older and older rocks until they have found a limit (temporary, we may hope) in the Trias. If mammalian bones should hereafter be discovered in the Coal Measures or Old Red Sandstone, the surprise would be, not that quadrupeds should have lived so long ago, but that their remains should have survived ten thousand chances of decay and obliteration.

The table, taken together with the corresponding summary of the Coal plants (p. 108), reveals the immense antiquity of the great types of animal and vegetable life. Nearly every existing class of animals is represented in the Carboniferous rocks; no Carboniferous fossil has rendered the creation of a new class necessary. As far as all the primary divisions, nearly all the classes and many of the orders are concerned, the Animal Kingdom of the Carboniferous period is the Animal Kingdom of to-day. It is no less requisite to a complete view that we should bear in mind that not a single well-understood species is known to descend from palæozoic to recent times. The specific difference is just as remarkable as the identity of type.

It has long been suspected that specific differences, and even those greater differences upon which classes and orders are founded, may be due to a continuous process of evolution. The common origin of various species now quite distinct may be said to be an established fact, and something is known as to the manner and the causes of their divergence. Palæontology has now and then

M

furnished important evidence by revealing a chain of
animal forms successively more and more altered in
the same direction as they descend from earlier to
later periods. The well-known cases of the Horses,
the Elephants, and the Crocodiles constitute perhaps the
nearest approach to direct proof of a process of evolution.
It seems natural to take a step in advance of this position,
to extend the operation of evolution to all organisms and
to all times, and to search the fossiliferous rocks for
the successive stages of development of perhaps one
primordial germ. Haeckel's 'History of Creation' may be
taken as the most elaborate of several attempts to realise
this programme. Here is traced the gradual differentiation of all the animal classes from certain Monera, which
at the beginning of the Laurentian period arose spon-
taneously by the association of simple combinations of
Carbon, Oxygen, Hydrogen, and Nitrogen. Echinoder-
mata and Mollusca appear later in the Laurentian;
Zoophytes, Arthropoda, and Vertebrata in the Cambrian.
Fishes first appear in the Silurian; the Ganoids diverge at
the beginning of the Devonian; the Teleostei in the
Jurassic. A great burst of new classes and orders marks
the beginning of the secondary period. At last the
Animal Kingdom of to-day is formed by the breaking up
of the tree into its ultimate ramifications.

We must not dwell in detail upon such a chart, though
the subject is in some ways tempting enough. It is
sufficient to point out that palæontology knows nothing of
all this—nothing of the derivation of the Vertebrates, for
example, nor even of the Fishes, nor even of the Elasmobranchs and Ganoids. Or take the Mollusca, palæontology
is altogether silent as to their origin, but it enables us to
trace all the five classes up to the Cambrian rocks, in which
the oldest known Mollusca occur. The chart agrees with

recorded facts just so far as the vertical lines of succession go; all the cross communications, all convergence of classes, in short, all that is distinctive in such a diagram, is purely imaginary, and much of it highly improbable.   The proof of this statement could not be given more clearly and fully than in Prof. Huxley's discourse on Geological Contemporaneity and Persistent Types of Life,[1] but the table of Carboniferous fossils, when well looked into, will do all that is necessary.   Even in the incalculably ancient Carboniferous age, the great majority of the principal types of animal life existed, just as distinct from each other, just as specialised in their structure then as now. It is conceivable that all animals have sprung from a common ancestor, but the fossiliferous rocks are appealed to in vain for the proof.   So remote is any possible ancestor of the entire Animal Kingdom that all known fossils are modern in comparison.   The main lines of descent run parallel throughout geological history ; there is convergence here and there, as we trace backwards the smaller threads, but the meeting-point of all the strands is, according to any human measure of past time, infinitely distant.

[1] *Lay Sermons,* p. 237 et seq. ; or *Quar. Journ. Geol. Soc.,* 1862.

CHAPTER V.

THE CHEMISTRY OF COAL.

WE have now to consider the nature of the chemical processes attending the conversion of woody matter into Coal.
Wood is a complicated mixture of many substances. A
very thin transverse section of it is seen under the microscope to consist of an aggregation of little cells more or
less filled with starch, sugar, gum, certain vegetable acids,
oils, resins and mineral substances, existing partly in the
solid state incrusting the cell-walls, and partly in solution
in the juices of the plant. The differences in woods of
different species or of the same wood at different periods
of growth depend on the different shapes and structures of
the cells, and on the nature and proportion of their contents. The chemical composition of the most important
of these contents, however dissimilar they may seem in
outward appearance and properties, is practically identical :
they consist, for the most part, of the elements carbon,
hydrogen and oxygen, in proportions varying within
comparatively narrow limits. One hundred parts of each
of the undermentioned bodies contain :

	Carbon	Hydrogen	Oxygen
Sugar	40·0	6·7	53·3
Starch	44·4	6·2	49·4
Gum	44·4	6·2	49·4
Cellulose	44·4	6·2	49·4
Lignose	51·7	6·2	42·1

The substance forming the main portion of woody tissue
is not, as is frequently supposed, *cellulose*: when freed

from soluble matter and all other foreign bodies, it has
the percentage composition :

Carbon	Hydrogen	Oxygen
48·5	6·2	45·3

This substance—the pure woody fibre—has apparently
the same composition whatever be the nature of the plant
from which it has been derived. It can be split up by
chemical treatment into sugar and a substance known as
lignose, and the lignose in its turn can be changed into
cellulose; indeed it is not at all improbable that these
transformations, which can be readily brought about in
the laboratory of the chemist, may under certain circumstances be effected within the living plant. In addition
to the carbon, hydrogen, and oxygen, the wood contains
small quantities of sulphur and nitrogen combined with
the organic matter, but the quantity of these substances
rarely exceeds two per cent.

Hence it follows that the chemical composition of the
organic matter of various plants, in spite of differences in
outward character, mode of growth, texture, hardness, &c.,
is nearly uniform, and analysis shows that the differences
between individuals of the same species are, in most cases,
very nearly as wide as the variations in the composition of
totally dissimilar plants. Subjoined are a number of
analyses of the wood of various forest-trees.

	Carbon	Hydrogen	Oxygen and Nitrogen	Mineral Matter or Ash
Oak	48·9	5·9	43·1	2·0
Birch	48·9	6·2	43·9	1·0
Beech	48·3	6·0	45·1	0·6
Cherry	48·9	6·5	44·3	0·3
Poplar	49·5	6·1	42·7	1·7
Willow	50·1	5·9	42·0	2·0
Pine	51·0	6·1	41·9	1·0
Fir	51·1	6·1	41·8	1·0
Mean composition of wood . .	49·6	6·1	43·1	1·2
Without ash.	50·2	6·2	43·6	—

Even the organic matter of ferns, lycopodiums, mosses

and horsetails, plants in all probability, as we have seen, more nearly allied to those from which coal has been mainly derived, has no very dissimilar composition from that of forest-trees. One hundred parts of air-dried club-moss (*Lycopodium clavatum*) for example contain :

	Whole plant				Without ash		
	C	H	O and N	Ash	C	H	O and N
Club-moss . . .	46·8	6·2	42·1	4·9	49·3	6·5	44·2

The fundamental fact that the ultimate composition of vegetable matter, even of plants of the most widely different character and type, is practically identical, affords strong presumptive evidence that the chemical nature of this vegetable matter has been the same throughout all time, and, therefore, that the chemical nature of the coal-plants was substantially the same as that of the plants of our own time ; a conclusion which tends to facilitate the consideration of the mode of their decay.

But there are certain parts of a plant the chemical composition of which frequently exhibits a very considerable divergence from that of the main bulk. This is most marked in the case of the seeds, or in the powdery granules which, as in the case of the spores of the club-moss, play the part of seed. One hundred parts of the spores of club-moss, for example, contain :

	Whole plant				Without ash		
	C	H	O and N	Ash	C	H	O and N
Spores of Club-moss .	61·5	8·4	27·7	2·4	63·0	8·6	28·4

This wide difference between the composition of vegetable matter in general and of the club-moss spores is mainly due to the large quantity of resinoid matter in the spores: it is specially interesting from the circumstance that the remains of such spores make up a considerable portion of the material of certain coals, as for example, of the Better Bed coal, which is so highly prized for the part it plays in the production of the most famous varieties of Yorkshire iron.

But even in the Better Bed coal, which is almost unique in its richness in spore-remains, there are portions which are unquestionably not derived from spores. The dull black, silky, fibrous, friable substance running between the laminæ or along the planes of bedding of a piece of coal, known as *Mineral Charcoal*, from its resemblance in outward appearance and chemical composition to ordinary charcoal, has in all probability been derived from leaves, or leaf-stalks, or possibly from bark and wood. It differs very widely from the adjoining bituminoid portions in ultimate composition, proportion of volatile matter, and amount and nature of ash, as the following analyses made on two typical Yorkshire coals, show :

	Mineral Charcoal		Adjoining Bituminoid portion	
	Better Bed	Haigh Moor	Better Bed	Haigh Moor
Volatile matter . .	18·9	23·2	33·7	40·8
Coke	74·8	67·3	63·1	56·4
Ash	6·3	9·5	2·2	2·8
Combustible matter : .				
Carbon . .	90·8	86·4	85·7	80·5
Hydrogen . .	3·6	3·9	5·4	5·5
Oxygen, &c. . .	5·6	9·7	8·9	14·0

The difference in the proportion of ash between the two portions is very remarkable, and is in conformity with their supposed mode of derivation. Of all portions of a plant the greatest amount of mineral matter is found in the leaves, leaf-stalks, and bark: as we have seen, the spores and the woody matter of the stems of trees leave comparatively little residue on incineration.

The general character of the ash of mineral charcoal seems to be tolerably uniform, judging from the following analyses of samples obtained from Haigh Moor and Better Bed coal.[1]

	Ash of Mineral Charcoal	
	Better Bed	Haigh Moor
Silica.	38·7	36·1
Alumina	33·8	28·7
Ferric Oxide	6·9	18·3
Lime	9·8	4·5
Magnesia	2·8	0·7
Sulphuric Acid	7·7	7·6
Chlorine, Alkalies, &c. . . .	0·3	4·1
	100·0	100·0

In the case of coals which furnish but little ash we are justified in assuming that the whole of the mineral matter has been contained in the plants from which the coal has been derived. The general nature of coal-ash seems to afford additional evidence of the origin of coal from lycopodiaceous plants. The ash of modern lycopods is characterised by the large proportion of alumina which they contain, due in all probability to the free acids present in the roots, by which they are enabled to dissolve this earth from the soil: indeed so large a quantity of acetate and malate of alumina does club-moss contain that an aqueous infusion of this plant was formerly used as a mordant in place of the ordinary 'red liquor' of the dyer. In no other plant except in Sphagnum has this peculiarity of absorbing alumina been noticed: direct observation has shown that the ash of trees such as the oak, birch, and pine, growing on the same soil as club-moss is perfectly free from this earth.

[1] From analyses made in the Laboratory of the Yorkshire College by Mr. Wood and Mr. A. H. Palethorpe.

The nature of the ash of different species of Lycopodium may be seen from the following analyses by Aderholdt.

	L. chamæcyparissus		L. clavatum
	With Spores	Without Spores	
Per cent. of Ash (dry plant) .	6·1	4·5	4·7
Alumina.	51·85	57·36	26·65
Iron and Manganese . .	2·96	2·75	4·83
Lime	5·41	4·81	7·96
Magnesia	3·97	3·21	6·51
Potash	13·18	11·92	24·19
Sulphuric Acid . . .	4·38	3·23	4·90
Alkaline Chlorides. . .	1·01	0·96	5·66
Phosphoric Acid . . .	3·63	2·71	5·36
Silica	13·60	12·96	13·94
	99·99	99·91	100·00

The most striking difference between the composition of the coal-ash and that of the lycopods is in the proportion of alkalies, which in the former ash is exceedingly small. But the alkalies are precisely those constituents which would be most readily removed by water during the process of decomposition. Dr. Vohl has made observations on the distribution of the inorganic matter of *Sphagnum commune* during decay under water, which have a direct bearing on this point. He analysed the ash of carefully selected moss, and subsequently allowed a portion to rot under water, and determined the nature and amount of the constituents in the water, and also those remaining in the plant.

	Original Moss	Decayed Residue	Aqueous Solution
Alkalies.	29·8	3·8	70·9
Ferric Oxide. . . .	6·4	13·4	0·2
Alumina	5·9	28·7	3·2
Lime	3·2	26·1	2·0
Magnesia	4·9	3·2	1·3
Sulphuric Acid . . .	4·3	6·0	3·2
Phosphoric Acid . . .	1·1	3·4	0·5
Silica	41·7	15·0	17·5

By the continued action of water charged with carbonic acid the amount of lime in the decayed residue would be gradually reduced, with the formation, eventually, of an ash very similar to that furnished by coal.

When wood is freely exposed to the air and rain it gradually gets rotten and is ultimately completely disintegrated. This change, even if it is not initiated by the air or by something in the air, is largely dependent on it: it is attended by the absorption of oxygen and the elimination of carbonic acid and water. A quantity of rotting wood which has been standing in a bottle in contact with oxygen in a warm place for a few days, will in this way furnish a considerable quantity of carbonic acid, which may be detected by passing the air from the bottle through a solution of quicklime in water; the clear lime-water will rapidly become turbid from the formation of the insoluble chalk, or carbonate of lime.

The greater part of the small quantity of nitrogen which wood contains appears to enter into the composition of certain substances in the cells, analogous in chemical nature to the white of egg, or *albumen*. These albuminoid bodies seem to initiate or to be primarily concerned in the rotting process—a kind of fermentative change is set up among them which gradually extends to the remaining constituents. Experience shows that young spongy woods or those of a succulent character are the most liable to decay: these contain comparatively large quantities of the nitrogenised bodies, and have a coarse cellular structure into which the air can readily penetrate. The process of decomposition is accompanied by a marked rise of temperature: a fact not unknown to certain little animals which are to be found in the decomposing mould: these not only enjoy the genial warmth, but find ample store of nutriment in the matter to which the change gives rise. The dark-brown mouldy substance into which the vegetable matter passes is known to chemists under the name of *Humus* or *Ulmin*. It is not a definite product but consists, for the most part, of certain imperfectly known acid substances characterised, among other properties, by their power of fixing the ammonia formed in the process of the alteration of the wood. Ammonia is one of the most important articles of plant food, and the value of mould in agriculture depends in great measure on its power of holding the ammonia in such form that it can be readily assimilated by plants.

But whatever may be the exact chemical nature of humus, analysis shows that it differs from vegetable matter in containing relatively more carbon and less oxygen and hydrogen than that substance.

	Carbon	Hydrogen	Oxygen and Nitrogen
Humus from Oak (Meyer) .	54·0	5·1	40·9
,, (Will) .	56·0	4·9	39·1
,, (Soubeiran)	55·3	4·8	39·9
,, Moss ,, .	54·0	4·6	41·4
Mean Composition. . .	54·8	4·8	40·4

By boiling sugar for a long time with dilute mineral acids, a substance very similar to, if not identical with humus is obtained. This fact is of interest as throwing possibly a little light on the chemical process by which humus is ordinarily derived from wood. Woody fibre and also cellulose itself can be made to yield sugar, and it is not at all improbable that in the decay of wood the humus may have been produced from the particular association of elements which furnishes the sugar.

The process of decomposition, however, when once set up, does not stop with the formation of the humus; the gradual elimination of the oxygen and the concentra-

tion of the carbon still go forward, and we see the result
of the alteration in the composition of the organic substance of *peat*, which is largely made up of altered humus
generally, although not altogether derived from the
decomposition of the vegetable matter of mosses. The
general chemical composition of peat may be seen from
the following analysis of samples obtained from various
localities.

	Carbon	Hydrogen	Oxygen and Nitrogen
Dartmoor, Devon . . .	59·7	5·9	34·4
Lewis, Scotland . . .	61·2	6·1	32·7
Bog of Allen, Ireland . .	61·0	5·8	33·2
Upper Shannon, Ireland. .	61·2	5·7	33·1

The external character and chemical composition of
peat vary with its age and with the position in the beds
in which it is found. The extreme variations are admirably illustrated in some of the extensive peat-bogs
for which the Sister Isle is famous, perhaps rather too
famous, for they cover some 3,000,000 of her acres, and
some of them are upwards of forty feet deep. The
upper layer is light in colour and soft and spongy in texture, and its vegetable origin is obvious even on the most
superficial inspection; as we go deeper down into the bog
the layers become darker coloured and are more strongly
pressed together, the vegetable structure becomes less
and less apparent until, as we near the base of one of the
deep bogs, the peat is almost black, has a density nearly
equal to that of coal, and requires rather careful scrutiny
to detect any organised structure. The chemical composition of the mass alters with this change in outward
character: the relative proportion of the oxygen steadily
diminishes whilst that of the carbon as steadily increases.
The altered vegetable matter now passes into a form

analogous to that met with in *Lignite*, a dark brown, somewhat friable substance of a compact woody texture.
Lignites are not very common with us, and where found
they are not much favoured as fuel on account of the
disagreeable smell they emit on burning.   Certain varieties
crumble to powder when thoroughly dried, and are used
as a pigment under the name of ' Cologne Earth.'   The
general composition of the combustible part of lignites
may be seen from the following analyses :

	Carbon	Hydrogen	Oxygen and Nitrogen
Lignite from Bovey, Devon . .	67·9	5·8	26·3
„ Cologne . . .	67·0	5·3	27·7

The term lignite may conveniently be restricted to
such varieties as show a distinct woody structure.   As
this disappears the substance passes into *Brown Coal*.
The brown coals proper contain a still larger percentage
of carbon and a smaller percentage of oxygen than the
true lignites.   They are not worked to any considerable
extent in this country, but on the continent and in
certain of our colonies they are very important fuels.

	Carbon	Hydrogen	Oxygen and Nitrogen
Brown Coal from Tasmania . .	71·9	5·6	22·5
„ Auckland . .	72·2	5·4	22·4
„ Hungary, slight fibrous structure	72·5	5·4	22·1
„ No fibrous structure Cubical fracture	74·9	5·2	19·9

From the brown coal we pass by insensible gradations
to the black coal or coal proper.   The change is attended
by an absolute loss of ligneous structure, by an increased
density, and a tolerably well developed cubical fracture.
When finely powdered, most black coals show more or less
of a brown colour in proportion corresponding to the

amount of oxygen which they contain. The following
are some analyses of well-known or typical coals :

	Carbon	Hydrogen	Oxygen and Nitrogen
Dudley, Staffordshire . . .	79·7	5·4	14·9
Wolverhampton, Staffordshire . .	79·9	5·2	14·9
St. Helen's, Lancashire . . .	79·9	5·5	14·6
Haigh Moor, Yorkshire . . .	80·5	5·5	14·0
Hartley, Northumberland . .	80·7	4·8	14·5
Low Moor, Yorkshire . . .	85·7	5·4	8·9
Newport, Wales	86·3	5·3	8·4
Risca, Wales	86·8	5·4	7·8
Wigan, Lancashire	87·3	5·5	7·2
Newcastle, Northumberland . .	87·9	5·3	6·8

These coals are examples of such as we usually burn
in our house-fires. Some of them, especially those from
the South Wales basin are valuable as Steam Coals.
House Coals are usually classed as *bituminous coals*, not a
very happy term, for there is nothing, strictly speaking,
of the nature of bitumen in them. Some coals when
thrown on the fire seem to fuse and swell up ; these are
technically known as *caking coals* in contradistinction to
the dry or freeburning coals which retain their shape but
tend to split up into columnar fragments. The cause of
this difference is not clearly made out : it is certainly not
dependent on ultimate composition, for two varieties of
coal occurring in the same bed may have the same proportions of elementary constituents and yet behave very
differently on heating. Many coals lose their power of
caking by long exposure to air; but on the other hand
the slack of non-caking coal may often be made to fuse
together into a compact mass if heated suddenly. The
principal caking coals are met with in the Newcastle coal-
field, whilst the free-burning coals are found largely in
Staffordshire and Scotland.

The blacker and harder varieties of Coal gradually
merge into the kind known as *Stone Coal* or *Anthracite.*

Anthracite has a brilliant lustre, is denser, harder, and more brittle than ordinary bituminous coal, and has a conchoidal fracture. It ignites with difficulty and gives out little flame on burning owing to the non-formation of volatile hydrocarbons. It is principally met with in South Wales, but our deposits of it are very insignificant when compared with those in Pennsylvania and in other parts of the American continent.

The manner in which anthracite has been formed has given rise to much discussion. The general opinion is that it is simply bituminous coal modified or altered by heat. Nevertheless the evidence on this point is far from being conclusive. Many coals tend to become anthracitic by simple exposure to air at ordinary temperatures. Better Bed Coal, for example, loses as much as 17 per cent. of volatile matter in two months' time (Miall). It has also been observed that coal in the vicinity of open faults, or immediately below a sandstone roof alters in texture, loses its cubical fracture, and is more highly carbonaceous than the ordinary bituminous varieties. Hence it must not be supposed that anthracite is necessarily the oldest because it is the most altered form of coal; the coal in the South Wales basin, for example, is of the same age throughout; nevertheless in the eastern part it is of the ordinary bituminous kind, and gradually becomes less bituminised as we pass westward, until it changes into anthracite.

The general composition of anthracite, or rather of that portion of it which burns away, may be seen from the following analyses.

	Carbon	Hydrogen	Oxygen and Nitrogen
Anthracite from S. Wales . .	93·5	3·4	3·1
„ Pennsylvania . .	94·9	2·5	2·6
„ Peru . . .	97·3	1·7	1·0

We have thus traced the passage of the woody matter
from the unaltered plants to the most altered form of coal,
and we see that the change is attended by the gradual
elimination of nearly all the constituents of the wood
except the carbon. The quantity of the hydrogen
apparently remains constant during the greater part of
the process of decomposition, but that this is only apparent
will be evident if we arrange the results as in the following table, in which the amount of carbon is considered
constant, viz. 100, and the proportion of the other constituents increased in the same ratio.

	Wt. of 1 solid cubic foot in lbs	Carbon	Hydrogen	Oxygen and Nitrogen
Wood, average . . .	30	100	12·3	86·8
Peat, „ . . .	50	100	9·7	54·7
Lignite, „ . . .	70	100	8·3	40·0
Brown Coal, average .	75	100	7·4	29·7
Bituminous Coal, average .	80	100	6·4	13·4
Anthracite, average . .	90	100	2·6	2·3

The numbers in the second column showing the
approximate weight of a solid cubic foot of the several
substances are instructive as indicating the joint effect of
compression and of the gradual destruction of cellular
structure in increasing the density of the product.

We have no means of knowing even approximately
what amount of woody fibre would be required to make
coal. Mohr has calculated that the transformation is
attended with a loss of 75 per cent. in weight, and that
when regard is had to the density of the two substances,
the volume of the coal is only about $\frac{1}{12}$ of the woody
matter from which it has been derived, but the data for
such computations are not very trustworthy.

Attempts have been made by Cagniard de la Tour,
Rivière, Daubrée, and others to imitate, by heat, the mode

in which coal has been produced from wood. According to Baroulier, vegetable matter, such as sawdust, twigs, stalks, and leaves, imbedded in moist clay and heated from 200° to 300° C for some time, yields a carbonised mass very similar to some varieties of coal.

We have now to consider the manner in which the oxygen and hydrogen have been removed from the vegetable matter. It has already been stated that wood decaying with free exposure to air and moisture evolves carbonic acid and water. But it is almost certain that the change of the vegetable matter to coal, even if commenced during free exposure to air, has not continued under these conditions; there can be little doubt that the greater portion of the rotting matter has been covered at a comparatively early stage of the decomposition by water and mud. The air, however, would not be entirely cut off by this submergence, for air is soluble to a slight extent in water. A gallon of rain water, for example, contains nearly seven cubic inches of gas in solution, half of which is oxygen: even sea-water taken from a depth of many thousand fathoms contains dissolved oxygen. The process of oxidation would still go on, therefore, under the water, but, of course, at a greatly diminished rate, owing to the relatively small supply of free oxygen. Nor would the decomposition stop when the initial supply was at an end, for there would be a constant interchange, by processes of diffusion between the gaseous products of the decomposition and the air above, at a rate depending on the depth to which the rotting matter was submerged. Masses lying some six or eight inches under water will decompose very rapidly, whilst at a depth of a couple of feet they are only slightly altered even after a year's time.

But the presence of water and the limited supply of free oxygen very materially modify the process of decomposition in another way. If a stick be thrust into the mixture of mud and rotting vegetable matter at the bottom of a ditch, a stream of gas-bubbles will be disengaged and these will occasionally ignite on the approach of a light with a slight explosion. The evolution of an inflammable gas from swampy districts has been observed from the earliest times; in some parts of the world these exhalations are so considerable and so continuous that flames of the gas have been burning there for very long periods; such a tract known as the *Field of Fire* occurs, for example, on the western shores of the Caspian Sea. Similar exhalations are sometimes noticed in the vicinity of brine-springs; the missionary Imbert relates that the Chinese have been in the habit for centuries of collecting the gas from the exhalations near certain of their salt-works, and employing it to boil down the brine and light up the manufactory. There is a certain kind of salt to be met with on the continent which crackles when thrown into water, owing to the forcible disruption of the grains by the strongly pent up inflammable gas as the crystals are thinned by solution. The same gas also occurs in the neighbourhood of the oil-wells in Pennsylvania, from which it is evolved at the rate of 1,000,000 cubic feet per hour: it is collected and conveyed in pipes for a distance of 35 miles to Pittsburg to be employed as a source of light and heat. This gas is sometimes found in the breath of sheep and cows, being derived from their food by a process of decomposition analogous to that occurring in the sodden matter of a bog. It has also been observed to be found, often to a very large amount, in the human body under certain special conditions. Sir William Gull has given the case of a drunkard who, after death, was so extraordinarily distended with the gas in all directions that, on puncturing the skin as many as fifteen or sixteen gas lights were procurable at once from his body, and these continued to burn until the whole of the gas was liberated.

The nature of this inflammable gas or *Marsh gas*, as it is not inappropriately termed, was indicated about a century ago by the Italian philosopher Volta, but it is to Dalton that we are indebted for our knowledge of its exact composition. Chemists regard it as containing one atom of Carbon and four atoms of Hydrogen. It can be artificially obtained in many ways. A very interesting mode of synthetically forming it has been pointed out by Sir Benjamin Brodie; it consists in subjecting a mixture of carbon monoxide and hydrogen to the action of the electric induction spark:

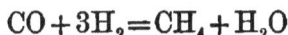

$$CO + 3H_2 = CH_4 + H_2O$$

As the carbon monoxide may be formed by the direct union of carbon and oxygen, we are able therefore to build up the marsh gas from its elements. As ordinarily prepared in the laboratory it is obtained by heating sodium acetate, a chemical compound of vinegar and soda, with lime and soda. Since the vinegar employed has been obtained from wood, we may say that this artificial marsh gas has been indirectly formed by the decomposition of woody fibre, as in nature. It is highly inflammable, burning with a bluish flame of feeble luminosity. When mixed with about nine or ten times its bulk of air it explodes with great violence on the approach of a light, or when it is otherwise heated to a sufficiently high temperature. When the proportion of air is increased to twelve volumes or diminished to six volumes, the mixture is neither inflammable nor explosive. The products of its combination or explosion are carbonic acid and water. Marsh gas has only half the density of air; hence it can

be poured upwards. If a bottle filled with the gas be
inverted under a vessel counterpoised on a balance, the
equilibrium will be at once disturbed, owing to the lighter
marsh-gas displacing the air (Fig. 30).

There is therefore this very essential difference between
the process of decay of woody matter in free air and under

FIG. 30.—Experiment for showing the relative density of
Marsh-gas and Air.

water, that whereas in the first case we have the formation
of carbonic acid and water; in the second, we have the
additional production of the marsh gas. In the one case
the carbon meeting with a liberal supply of oxygen is

eliminated entirely as carbonic acid; in the other where
the supply of oxygen is but small only a portion of the
escaping carbon is oxidized, the rest is evolved in union
with hydrogen, partly derived, it may be, from the woody
matter itself and partly from the associated water. It is
probable, as will be shown hereafter, that under certain
circumstances other gaseous products are produced, but in
quantity too small to affect the general result.

Nevertheless it would be quite futile to attempt to
formulate by chemical equations the process of transformation of woody matter into coal. Such equations have,
indeed, been devised, but with the data at present in our
possession they are perfectly valueless, and indeed posi-
tively misleading.

A further proof that marsh gas is actually produced in
the formation of coal is afforded by the fact that it is often
met with in the atmosphere of coal-mines; it constitutes
the *fire-damp* of the miner. The fire-damp oozes imper-
ceptibly from the face of the coal or escapes from cracks
or fissures in jets, known as 'blowers.' Occasionally the
mass of the coal is so highly saturated with the gas that
it continues to escape from it for some time after it has
been raised; a fact well known to persons engaged in
storing or shipping coal; it not unfrequently happens that
the coals in a ship's hold, for example, give out such a
large quantity of this occluded gas that the air in the
immediate neighbourhood becomes highly explosive.

In a vacuum, especially at a gentle heat, this occluded
gas is readily disengaged from the pores of the coal. The
flask standing in the water-bath (Fig. 31) is filled with
pieces of the freshly-raised coal: on exhausting the apparatus by the aid of the Sprengel pump and then raising the
temperature of the bath to the boiling-point the gas is
evolved and may be collected in the tube previously filled

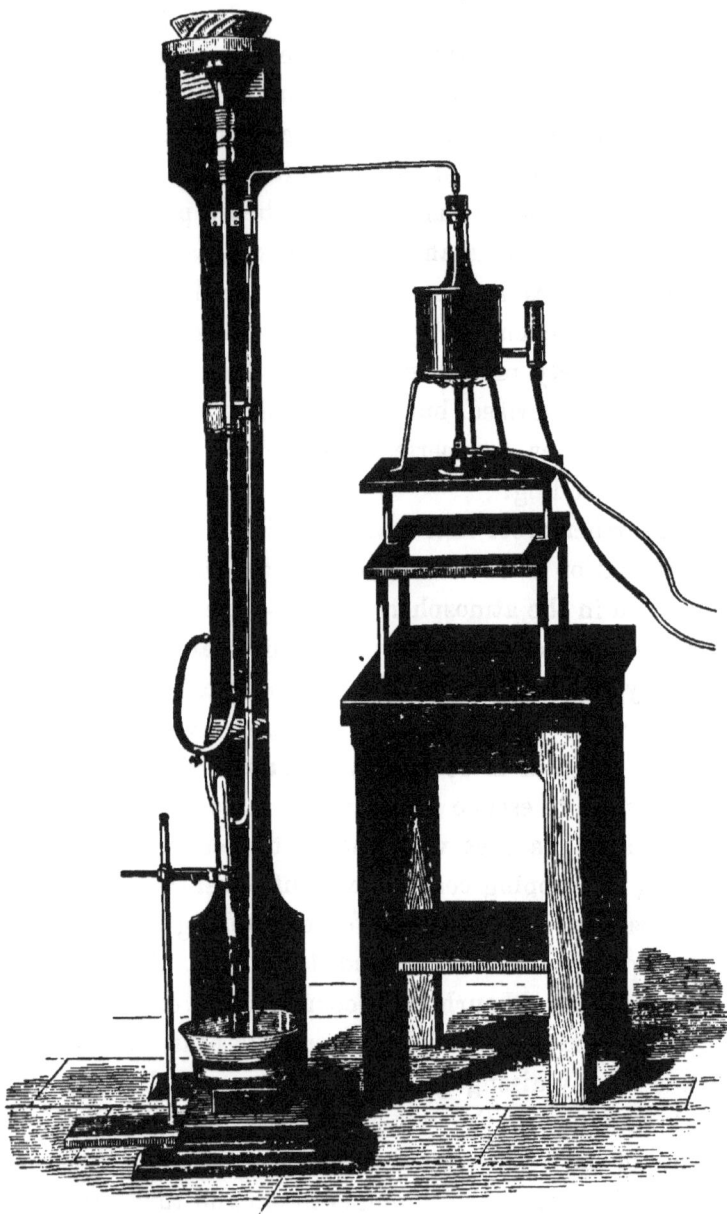

FIG. 31.—Apparatus for the collection of gases occluded in Coal.

with mercury and placed over the end of the fall-tube of the pump.

A study of the nature of the gases contained in the pores of different varieties of coal is calculated to throw considerable light on the mode of their formation. Quite recently Mr. J. W. Thomas has published a series of analyses of the gases obtained by exhaustion *in vacuo*. He finds that the bituminous coals of the South Wales basin, which contains one third of the available coal of Great Britain, contain very little gas, and that little is almost exclusively carbonic acid. The atmosphere of the workings of these mines is almost free from marsh gas, the chief gaseous impurity being the carbonic acid. With the increase in the carbon, both the total volume of the occluded gas and the relative proportion of the marsh-gas become larger; this is especially the case in the hard compact steam coals. Anthracites yield by far the largest volume of gas, and this consists almost entirely of marsh-gas; 1 lb. of some anthracites will give off nearly a couple of gallons of gas. The enormous quantity of gas retained by anthracites is due to their compactness and dense structure, which again arises from the great pressure to which they have been exposed. On the other hand the steam coals, and especially the bituminous coals, are less coherent and more porous, and the gases much more easily make their escape. Enormous volumes of gas rush out from the working face of some of the deep steam-coals whilst comparatively little escapes from that of anthracites.

The gases occluded in cannel are distinguished from those met with in the other coals by the presence of the hydrocarbon *Ethane*, which differs from marsh-gas in containing more carbon and hydrogen, its chemical formula being C_2H_6. It burns with a bright luminous flame, and doubtless contributes to the illuminating power of the gas obtained by heating cannel on a large scale. *Jet*, which is a species of cannel, affording a large quantity of illuminating gas when heated, contains a still larger proportion of occluded hydrocarbons of even higher molecular weight than ethane.

The gas in brown coal and lignite consists almost entirely of carbonic acid mixed with a small quantity of carbonic oxide and nitrogen. The presence of the carbonic oxide is especially remarkable, as it has not been detected in any other carboniferous exhalation.

Some of the results of Mr. Thomas's analyses are seen in the following table :

	c.c. of gas from 100 grams. evolved at 100°	Percentage Composition of Gas					
		CO_2	CO	CH_4	C_nH_n	O	N
Lignite, Bovey . . .	114·3	96·74	2·80	—	—	—	0·46
Cannel, Wigan . . .	350·6	9·05	—	77·19	7·80	—	5·96
Jet, Whitby . . .	30·2	10·93	—	C_4H_{10} { 86·90	—	2·17
Bituminous Coal, S. Wales	55·9	36·42	—	—	—	0·80	62·78
Semi-bituminous „ .	73·6	12·34	—	72·51	—	0·64	14·51
Steam Coal, S. Wales .	218·4	5·46	—	84·22	—	0·44	9·88
Anthracite, „ . .	555·5	2·62	—	93·13	—	—	4·25

Fire-damp constitutes one of the most formidable of the evils with which the coal-miner has to contend; in some pits the amount evolved is so large that it is only by special precautions that the proportion in the air is kept below the explosive point. It is only during the last sixty years that a mode has been discovered by which these so-called 'fiery' pits can be worked with comparative safety. It was at all times highly dangerous and often absolutely impossible to venture into such places with a naked flame, and the only light by which the miner could work was furnished by the *steel mill*, an apparatus which if we may judge from a reference in 'An Account of the Damp in a

Coal Pit of Sir James Lowther sunk within Twenty yards of the Sea,' in the 'Philosophical Transactions' for 1733, seems to have been brought into use during the early part of that century.   It consisted of a steel disc (Fig. 32) made to revolve rapidly against a piece of flint so as to produce a shower of sparks: a contrivance both clumsy and unsafe.   Sir Humphry Davy the chemist, and George Stephenson the engineer, independently and almost simultaneously, hit upon a method of so protecting a candle or oil flame that it could be brought into an explosive atmosphere with impunity.   The lamps depend for their safety upon the fact discovered by Davy that fire-

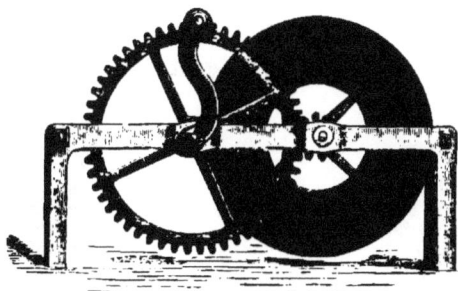

FIG. 32.—Steel Mill.

damp requires a very high temperature to cause it to burn ; if when the gas is burning its temperature is rapidly lowered the flame is extinguished.   This principle admits of many experimental illustrations.   If a piece of wire gauze not exceeding a certain coarseness be placed upon a flame (Fig. 33), the flame does not pass through the gauze for the reason that the metal cools the gases below the ignition point; if however by prolonged contact with the flame the gauze is allowed to get red-hot it will no longer exert this cooling action and hence the flame will traverse it.   Figs. 34, 35, show the Davy lamp.   The oil flame burns within the gauze cylinder, and the fire-damp which finds its way through the meshes may also burn within the cylinder, but the flame is powerless to pass through the gauze on account of the cooling power of the cold metal.

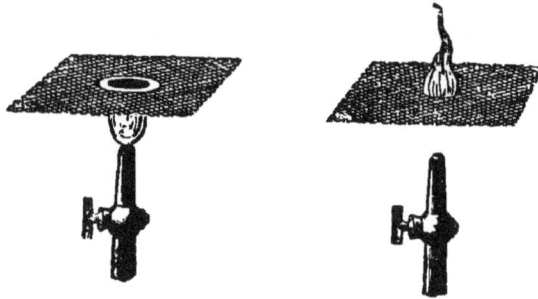

FIG. 33.—Experiment to show that flame will not pass through gauze.

If, however, the wire by some chance gets very hot, the flame may pass through and ignite the explosive mixture

FIG. 34.—Davy Lamp (Section).   FIG. 35.—Davy Lamp (Elevation).

on the outside, exactly as the coal-gas flame passed through the red-hot gauze.

Here are a number of figures taken from Davy's account of his lamp; they are interesting as showing the successive steps of his great invention.

Fig. 36 represents one of the first lamps made by

Davy; it is simply a horn lantern with air-tight sides. The air necessary to feed the flame passes through a number of concentric metallic cylinders at the bottom, and the products of combustion pass out through a simi-

FIG. 36.   FIG. 37.   FIG. 38.

 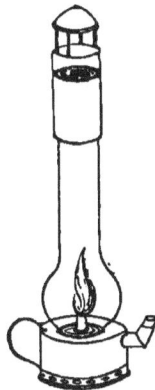 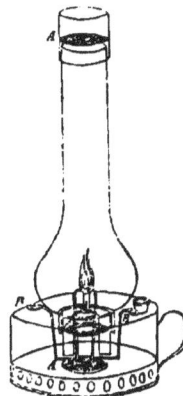

FIG. 39.   FIG 4 .

 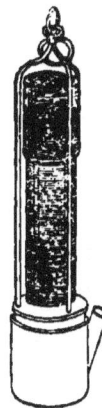

Successive steps in the invention of the Davy Lamp.

lar series at the top. An explosion may occur within the lamp, but the outward passage of the flame is prevented by the cooling action of the metallic rings.

In Fig. 37 the horn sides are replaced by a glass chimney. In Fig. 38, wire gauze is substituted for the metallic cylinders. In Fig. 39 the fragile glass chimney is replaced by a cylinder of wire gauze, and lastly Fig. 40 shows the lamp as Davy left it.

Although experience has shown that a considerable proportion of fire-damp may be breathed with impunity, the safety lamp was not intended to allow men to work in an atmosphere which is constantly at the explosive point; and in all properly-regulated collieries it is understood that under no circumstances must work be conducted in places where ' fire ' has been shown to be present, until it is so far diluted with air as to be inexplosive. The safety lamp as ordinarily used by the miner ought simply to be regarded as a precaution against a local outburst of gas, or an interruption to the ventilating current.

The Davy lamp was unquestionably a marvellous gain when compared with the clumsy steel-mill, and even the Davy itself has during the last few years been very greatly improved upon, as regards security and illuminating power. Nevertheless the lighting of our coal-mines is hardly so satisfactory as to preclude the possibility of further improvement. Attempts have been made to illuminate by electricity but at present with only partial success, although the measure of this success according to some authorities has been sufficient to justify the belief that in this agent we have the mode of illumination of the future.

When the electric spark is passed through a rarefied gas, the particles of the gas are heated to incandescence and we have the effect of light. On this principle MM. Dumas and Benoit have conducted the electric lamp seen in Fig. 41. A discharge from an electro-magnetic apparatus contained in the box passes through the spiral tube and heats the small quantity of gas which it contains sufficiently high to cause it to give out a considerable amount of light. This apparatus, although meriting

notice as an ingenious application of a very beautiful
principle, is far too costly and fragile to be of any general
practical value, although in certain occasional cases, as in
the exploration of workings so highly charged with fire- .
damp that the ordinary safety-lamp becomes useless, it
might be of service. It has not unfrequently happened
that after a heavy explosion all the work necessary to

FIG. 41.—Dumas and Benoit's Electric Lamp.

restore the ventilation has had to be carried on in dark-
ness and has consequently been greatly prolonged; under
such circumstances a lamp on this principle might be of
considerable benefit.

During the last few years increased attention has been
paid to the subject of the causes of colliery explosions in
general: thus, among other things, they have been shown

to be connected with extreme meteorological changes—a rapid fall in the barometer or a rapid rise in temperature tending to bring about an increased liberation of fire-damp; and warnings are now issued by the Meteorological Department to colliery managers on the occasion of any great atmospheric disturbance. Practical men sometimes complain of the tardiness of the ordinary barometer in indicating change, and allege that it often follows the effect it is required to predict. Possibly the glycerine barometer of Mr. J. B. Jordan, which was exhibited in the

FIG. 42.—Ansell's Indicator.

Loan Collection of Scientific Apparatus at South Ken-sington last summer, might meet this objection: it indi-cates small changes of pressure by large oscillations of a fluid column exerting practically no vapour tension.

Various instruments have been devised for detecting the presence of fire-damp in a mine: one of the most ingenious of these is known as Ansell's Indicator, and is represented in Figs. 42. It is based on the high diffusive power of marsh gas as compared with ordinary air. It has already been said that marsh gas is much lighter than air: and there is an intimate physical con-nection between the specific gravity of a gas and its power of diffusion; the lightest gases diffuse most rapidly. Supposing we take a closed porcelain vessel filled with air and sufficiently porous to allow gas to flow through it; if

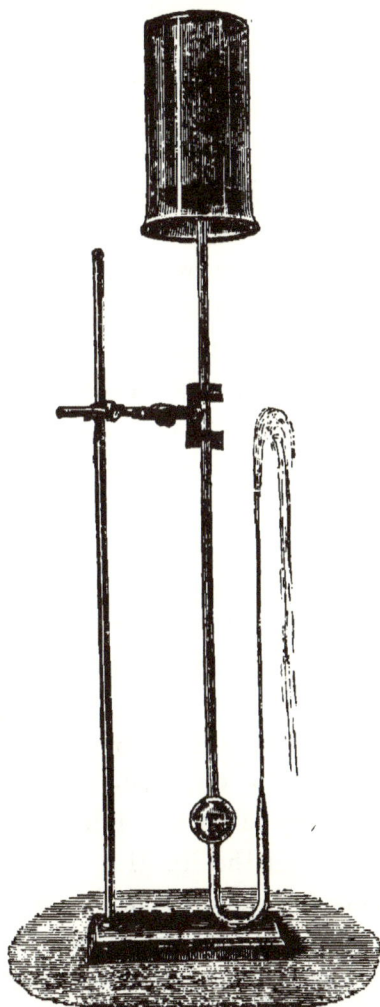

FIG. 43.—Apparatus for illustrating the Diffusion of Gases.
From Roscoe and Schorlemmer's ' Chemistry.'

standing in air, the outside air will go through to the inside, and an equal volume of air from the inside will pass outwards through the pores; there will be a constant

interchange of the air from within to without and *vice
versá*, and since the volumes passing in and out in a given
time are identical, no difference of pressure will be perceived within the porous cell.   If the cell, closed at the top,
be connected with a long bent tube containing some
coloured water (Fig. 43), no change in the level of the
water in the tube will be apparent so long as the gases
within and without are of the same kind or have the same
rate of diffusion.   If a jar filled with marsh gas be brought
over the cell an immediate increase in pressure within the
cell is observed, for the reason that 132 cubic inches of the
marsh gas enter through the pores in the same time that
100 cubic inches of air pass out; hence the coloured
water is driven out of the tube in the form of a fountain.

The upper part of Mr. Ansell's instrument consists of a
porous porcelain plate *m* cemented into an iron vessel containing air, and partially filled with mercury.   On bringing
the apparatus into an atmosphere charged with fire-damp,
the gas passes through the porous plate more rapidly than
the air can pass out, the mercury is accordingly driven
against the wire *f*, and makes an electrical connection
with an alarm-bell.

It has been clearly shown by Mr. Galloway that a
sound wave of a certain amplitude of vibration will drive
the flame of a Davy lamp through the wire gauze, and
may thus bring about the ignition of an explosive mixture on the outside; and he has also proved that the
presence of coal-dust in the atmosphere of a mine
largely contributes to its explosive character.   A sad experience has shown how premature was Buddle's exulting
exclamation, ' At last we have subdued this monster!' as
he saw how the first Davy lamp hanging in a fiery place
grew red-hot without inflaming the gas.   The remote
causes of colliery explosions are frequently totally inex-

plicable according to our present knowledge, or are, at most, bare matters of conjecture ; recent observations show that agencies like those above mentioned, some of which as yet are not very clearly understood, are far more frequently at work than has been supposed.

Coal-mining, despite every care, is, and must in all probability continue to be, a hazardous occupation: our yearly supply of coal is attended with the sacrifice of a thousand lives, about a fifth of which loss is, on the average, traceable to fire-damp. Nevertheless the proportional loss of life in our collieries is much less than in other coal-producing countries. In Germany the loss of life in 1874 was 1 in 331, with a produce of 65,000 tons for each death ; during the same period our loss was 1 in 510, with an out-put of 133,000 tons for each death.

CHAPTER VI.

THE CHEMISTRY OF COAL—*continued*.

OF the hundred and odd million tons of coal which we in this country burn in the course of a year, about 20,000,000 tons are thrown on to our house-fires; 30,000,000 tons find their way into our blast furnaces, or are otherwise used in the smelting and manufacture of metals; about 48,000,000 are burnt under steam boilers; 6,000,000 are used in gas-making, whilst the remainder is consumed in potteries, glass works, brick and lime kilns, chemical works, &c., &c.

A piece of coal when thrown on the fire very quickly shows signs of chemical change. As it gets heated little jets of smoke issue from it which ultimately burst into a bright smoky flame; after a time the smoke and the flame disappear, the mass becomes uniformly red-hot, and gradually burns away, partly to invisible gases which pass up the chimney and partly to ashes which remain behind in the grate. The products of the action of heat upon coal are very numerous; we shall have occasion to mention some of them in treating of the manufacture of coal-gas. But when coal burns in contact with an unlimited supply of air, these products are eventually resolved into three or four substances only; the carbon is converted into the two oxides, carbon dioxide and carbon monoxide; the greater part of the hydrogen combines with oxygen to form water, whilst the remainder unites with some of the nitrogen in the coal to form ammonia, the rest of the nitrogen escaping in the free state.

It is easy to demonstrate the fact that burning coal forms carbon dioxide and water. If a quantity of the fuel is mixed with some substance which readily parts with oxygen, such as oxide of copper, and the mixture is heated in a tube through which a current of air or oxygen is driven, the water which is formed by the combustion of the hydrogen may be condensed in a little receiver attached to the tube; whilst the carbonic acid formed by the union of the carbon and oxygen may be detected by passing the escaping gases through lime-water, which quickly becomes turbid from the formation of chalk. By weighing the carbon dioxide and water thus formed, the chemist can calculate from the known composition of these substances, the proportion of carbon and hydrogen contained in a given weight of the coal. It is on this principle that the ultimate analysis of the fuel is made.

The blue lambent flame seen playing over the red-hot coal in a clear fire is due to the carbon monoxide combining with oxygen to form carbon dioxide. Very little, if any, of this gas escapes unburnt up the chimney under ordinary circumstances to mix with the outer air, and fortunately so, for this carbon monoxide is one of the most poisonous substances known to the chemist.

Dr. Angus Smith has analysed the gases present in a common house fire with the following results:

	Carbon dioxide	Carbon monoxide	Oxygen	Nitrogen
Gas from the clear fire, below .	16·10	—	4·95	78·95
Gas from a heap of glowing Coal	18·17	2·48	—	79·35
Gas from the upper part of the fire, one inch below surface	20·80	0·99	—	79·21

It will be noticed that it is only in the lower part of

the fire that free oxygen can be detected: here no carbon
monoxide is found; the two gases in fact cannot co-exist
at the high temperature of the fire. As the carbon
dioxide passes over the glowing coal a portion of it appears to take up an additional quantity of carbon to form
carbon monoxide: thus, $CO_2 + C = 2\ CO$.

The greater portion of the nitrogen of course comes
from the air: the water and the very small quantity of
ammonia which may escape decomposition are not given
in the analysis.

Whenever coal is badly and wastefully burnt, as it is
in the production of dense black smoke, the poisonous
carbon monoxide is partially unburnt and escapes up the
chimney. This fact is proved by analyses of the gases
formed in smoke taken from the chimney of a large works,
for which we are also indebted to Dr. Angus Smith.

	CO_2	CO	O	N
Dense black smoke from bottom of Chimney . . .	7·07	6·00	7·92	79·01
Dense black smoke from opening	6·17	1·55	12·22	79·93
Common brown smoke . . .	5·05	none	14·41	80·54

Here then we have an additional reason why the
plague of black smoke which affects so many of our towns
should be stayed. The soot not only soils our persons
and clothes, and dirties our buildings and furniture, but
its pernicious concomitant, the carbonic oxide, impairs
the vitalising action of our air.

But there is another constituent of coal-smoke which,
although too little in quantity to be taken cognisance of
in the small bulk of chimney gas usually taken for analysis, is in reality evolved in considerable amount when we
have regard to the enormous consumption of coal in
our larger towns. This baneful substance is the sulphur

which exists in the coal-smoke as *sulphur dioxide*, the
name given by chemists to the pungent, choking gas
given off from burning brimstone. Everybody must have
noticed the bright, brass-like particles of *iron pyrites*
in coal. This substance has been formed in the course
of the decomposition of the woody matter. The vegetable
matter and the water which saturates it contain sulphates
which in contact with the decaying organic substance are
converted into sulphides ; these re-act upon the iron also
present in the plant or in the water and form the pyrites
or iron sulphide. The presence of sulphides formed by
the mutual action of organic matter and sulphates may
be detected at the mouth of any large tidal river where
the decomposing organic matter brought down by the
stream re-acts upon the sulphates contained in the sea-
water ; the same action may be observed to occur in the
black mud at the bottom of ditches and cess-pools, the
colour of which is mainly due to the intermixture of
finely divided sulphide of iron. The amount of iron-
pyrites in coal of good quality should not exceed 3 per
cent., but some coals, as the *stinking coals* of South Staffordshire, contain it in much larger quantity—so large
indeed that they are almost valueless as fuels. When
the coal is burnt the sulphur is mainly converted into
sulphur dioxide, which escapes up the chimney, whilst the
iron is left behind in the ashes, to which it gives a red
colour. On the average a cubic foot of coal-smoke contains about half a grain of the sulphur oxide. This is eventually converted into sulphuric acid or oil of vitriol, and is
brought down by the rain, and thus effects the corrosion
and disintegration of stone and brick-work by dissolving
out the lime, magnesia, and alkalies which these materials
contain. Dr. Angus Smith who has given great attention
to the subject of the nature and effect of the impurities in

air,[1] has determined the acidity of a large number of specimens of rain-water collected in our larger towns with the following results :

Amount of free Sulphuric Acid in the Rain-water of Towns.

	Grains per gallon	Parts per million
London, 1869	0·33	4·71
Manchester, 1869 and 1870 . . .	0·89	12·43
Liverpool	0·99	14·14
Glasgow	1·297	18·53

The heat we get out of coal is the result of the chemical combination of the oxygen of the air with the carbon and hydrogen of the fuel. The development of chemical action is in the vast majority of cases attended with the production of heat. The warmth of rotting vegetable matter, the slaking of lime, the spontaneous combustion of greasy wool are familiar instances of this formation of heat by chemical action. The very warmth of our bodies is one of the most striking examples of the fact. Many experimental illustrations might be given of the intimate relation existing between heat and chemical action. The Döbereiner lamp (Fig 44) is perhaps a perfect exemplification of their interdependence. This ingenious lamp came into use about the year 1823 as a substitute for the old flint and steel and tinder box, some years before the introduction of the speedy lucifer. A stream of hydrogen, generated by the action of dilute sulphuric acid on the rod of zinc contained in the inner vessel, issues from the mouth of the little cross-legged Kobold when his extinguisher is raised, and is directed on the small mass of spongy platinum hanging from the thin platinum wire. The porous platinum absorbs or occludes oxygen from the air in some mysterious way and with

[1] For further details, see Dr. Smith's *Air and Rain: the beginnings of a Chemical Climatology.* Longmans.

this oxygen the issuing hydrogen combines with the development of so much heat that the mass of the metal is raised to a red heat and ignites the stream of gas.

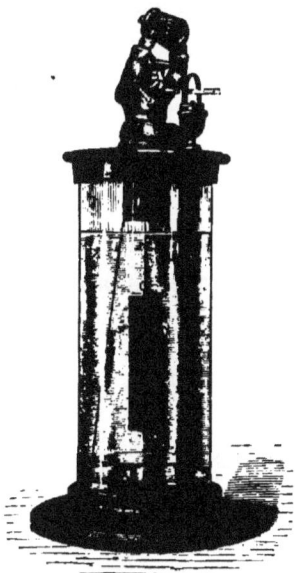

FIG. 44.—Döbereiner Lamp.

The amount of heat generated by a known amount of hydrogen or of carbon burning in the air is a fixed and definite quantity, so that if we know the chemical composition of coal we can calculate the amount of heat which it is theoretically capable of yielding. How this may be done, and how far the calculated quantity differs from that actually realised in practice, will be explained subsequently.

When coal is heated in closed vessels, that is under such conditions that the air cannot act upon the products at a high temperature, the result is very different to that observed in the open fire. The coal evolves a large number of substances, liquids, solids, and gases, and yields a residue or *coke* relatively richer in carbon than the original coal. The liquid products separate into two layers; the lower one is *tar*, whilst the upper one consists of a watery solution of various ammoniacal salts, chiefly the sulphide and carbonate. The relative proportion of the main products varies with different coals, and in the same coal with the temperature to which it is heated: at a low heat the tar is evolved in largest quantity whilst at a high temperature gases are mainly formed.

In the following table are the names and chemical formulæ of some of the substances obtained by the destructive distillation of coal:

Hydrogen, H

Nitrogen, N

Methane, CH_4

Ethane, C_2H_6

Propane, C_3H_8

Butane, C_4H_{10}

Pentane, C_5H_{12}
    &c. &c.

Solid paraffins C_nH_{2n+2}

Ethene, C_2H_4

Propene, C_3H_6

Butene, C_4H_8

Pentene, C_5H_{10}

And other members} of the C_nH_{2n} series}

Ethine, C_2H_2

And other members} of the C_nH_{2n-2} series}

Benzene, C_6H_6

Toluene, C_7H_8

Iso-xylene, C_8H_{10}

Methyl-toluene, C_8H_{10}

Cumene, C_9H_{12}

Pseudo-cumene, C_9H_{12}

Mesitylene, C_9H_{12}

Cymene, $C_{10}H_{14}$

Napthalene, $C_{10}H_8$

Styrolene, C_8H_8

Metastyrolene $(C_8H_8)_n$

Phenanthrene, $C_{14}H_{10}$

Anthracene, $C_{14}H_{10}$

Pyrene, $C_{16}H_{10}$

Chrysene, $C_{18}H_{12}$

Idrialene, $C_{22}H_{14}$

Ammonia, NH_3

Aniline, C_6H_7N

Toluidine, C_7H_9N
    &c., &c.

Pyrrol, C_4H_5N

Pyridine, C_5H_5N

Picoline, C_6H_7N

Lutidine, C_7H_9N

Collidine, $C_8H_{11}N$

Parvoline, $C_9H_{13}N$

Coridine, $C_{10}H_{15}N$

Rubidine, $C_{11}H_{17}N$

Viridine, $C_{12}H_{19}N$

Leucoline, C_9H_7N

Lepidine, $C_{10}H_9N$

Cryptidine, $C_{11}H_{11}N$

Phenol, C_6H_6O

Cresol, C_7H_8O

Phlorol, $C_8H_{10}O$

Carbon monoxide, CO

Carbon dioxide, CO_2

Sulphur dioxide, SO_2

Sulphydric Acid, H_2S

Carbon bisulphide, CS_2

Sulphocyanic Acid CNSH

&c., &c.

More than a hundred and fifty years ago Dr. Stephen Hales in his ' Statical Essays ' told us how he obtained 180 cubic inches of inflammable gas, weighing 51 grains, by distilling 120 grains of Newcastle coal in an earthen retort, and since that time every schoolboy with a taste for practical chemistry has become familiar with this experiment by heating coal in a clay tobacco-pipe closed with clay.

In the latter part of the seventeenth century, the Rev. Dr. Clayton observing the issue of inflammable gas from a coal seam in the neighbourhood of Wigan, which exhalations had previously been described by Mr. Shirley in the 'Philosophical Transactions' for 1667, was induced to try if a similar gas could not be obtained by heating coal in retorts over an open fire. He noticed the formation of 'a black oil' and 'a spirit' which he tells us could in nowise condense, but which blew up a bladder 'flatted and void of air' fixed to the pipe of the receiver almost as fast as a man could blow it up with his mouth. When he wished to amuse his friends he would take one of the inflated bladders and pricking a hole with a pin and compressing gently the bladder near the flame of a candle till it once took fire, it would then continue flaming till all the spirit was compressed out of the bladder.

Dr. Watson in his 'Chemical Essays,' published in 1767, describes how he obtained 28 ounces of gas or air, 12 ounces of water and tar, and 56 ounces of coke by the distillation of 96 ounces of Newcastle pit coal. 'The air which issued with great violence from the retort was inflammable, not only at its first exit from the distillatory vessel, but after it had been made to pass through two high bended glass tubes and three large vessels of water.' About thirty years later Mr. William Murdoch, a Scotchman in the employ of the great engineers Boulton and Watt, turned these observations to account, and in 1802 he practically demonstrated the value of coal gas as an illumina-

ting agent by lighting up the famous Soho factory on the
occasion of the rejoicings following the Peace of Amiens.
. An eye-witness thus describes the event. 'The
illumination of Soho works on this occasion was one of
extraordinary splendour. The whole front of that extensive range of buildings was ornamented with a great
variety of devices that admirably displayed many of the
varied forms of which gas-light is susceptible. This
luminous spectacle was as novel as it was astonishing;
and Birmingham poured forth its numerous population to
gaze at and to admire this wonderful display of the
combined effects of science and art.'

In the same year a Frenchman, Lebon, apparently in
ignorance of Murdoch's doings, lighted a house in Paris by
means of gas. The noise of Murdoch's invention gradually spread : other manufacturing establishments followed
the example of the 'manufactory of power' at Birmingham,
and in 1812 the first public gas company—the London and
Westminster Chartered Gas-light and Coke Company—
was incorporated by Act of Parliament. This concern,
now the largest of its kind in the world, owes its origin as
much to the eccentricities as to the energy of a wonderful
compound of credulity, charlatanry, and common-sense,
known as Winsor. This man, a German, belonging to
Frankfort, hearing of Lebon's success, hurried to Paris,
but even the ' offer of 100 Louis d'or' failed to obtain for
him the knowledge of the secret. By some means, however, he eventually procured the desired information, but
his countrymen either ridiculed or were stolidly indifferent
to his plausible plans for dispelling the Cimmerian gloom
of their towns, so he turned, with the perfect faith of his
kind, to the British public. ' The thought,' he wrote, ' of
introducing the discovery for the advantage of the British
realm struck me like an electric shock.' 1804 found him

lecturing on ' Gaslight ' at the Lyceum Theatre, but unfortunately to the obduracy of the English mind were added
the mysteries of the English tongue, and Winsor knowing
little of our language was compelled to rely on the eloquence
of a coadjutor, who, seemingly so far failed to catch the
all-absorbing enthusiasm of his chief, as from time to time
to forget his engagement, or inconveniently to leave the
manuscript at home.   But no amount of ridicule or ill-suc-
cess could damp the ardour of Winsor, and no projector
of Joint Stock Companies ever worked more perseveringly
than he did to make known the marvels and virtues of
the new light.   Our great-grandfathers had a weakness
for illuminations; hence the information that these
' may be carried on to the utmost extent of beauty and
variegated fancy by this docile flame, which will ply in all
forms, submit to instant changes, ascend in columns to
the clouds, descend in showers from the trees and walls,
arise from the water, and even in the same pipe with a
playing fountain.   The constant varying of flames in
rooms and gardens, between flaming pyramids, festoons,
garlands, roses, palm-twigs, suns, stars, urns, torches,
flambeaux, &c., afford to the spectator an extraordinary
and most delightful sight, cherish the soul, and create good
humour by uniting convenience, utility, and pleasure.'
When prose failed other means were used ; and the age
which found pleasure in contemplating the unhappy union -
of Calliope with the genius of Chemistry, in Dr. Parkes'
' Chemical Catechism,' was thus invoked :

Must Britons be condemned for ever to wallow
In filthy soot, noxious smoke, train oil, and tallow,
And their poisonous fumes for ever to swallow ?
For with sparky soots, snuffs, and vapours men have constant strife ;
Those who are not burned to death are smothered during life.[1]

---

[1] See *A Treatise on the Science and Practice of the Manufacture and Distribution of Coal Gas.*   W. B. King, London.

Winsor succeeded in 1807 in lighting Pall Mall, but Parliament at first refused to incorporate his National Light and Heat Company. The new light had to struggle for a time against prejudice and misrepresentation; Clegg, the energetic manager of the new company, and to whom its success was undoubtedly due, had to light the gas-lamps on Westminster bridge himself, such was the dread which the inflammable gas inspired. When it was intro-duced into the House of Commons, the architect insisted that the mains should be placed a certain distance from the walls to prevent (as he said) the heat from the pipes setting fire to the building. Sir Joseph Banks and a Committee of the Royal Society visited the company's works to report on the expediency of allowing large volumes of gas to be stored, and were amazed to see Clegg drive a pickaxe into a gas-holder and light the issuing gas with impunity. But the opposition was comparatively short-lived, and now within this short intervening space of a couple of genera-tions, the manufacture of coal gas has extended to nearly every country in the world, and there is scarcely a city of importance which is not lighted by it.

Coal-gas is made by heating coal or cannel, which is the special form of coal most valued for the purpose on account of the high quality of the gas it produces, in cylindrical fire-clay retorts A (Figs. 45 and 46), ranged horizontally in a furnace. The gases pass up the vertical pipes H fixed to the front of the retorts and are sent into the *hydraulic main* G, G, in which the more readily lique-fiable products condense. The height of the liquid in the hydraulic main is so regulated that the pipe *n* dips to the depth of three or four inches below the level; this prevents a rush of air into the main when the ends of a retort are removed for recharging. As the liquid accumulates in the hydraulic main it flows over into the tar-well, and the

gas passes on through a series of vertical iron tubes
termed *refrigerators* or *condensers,* two of which are seen
in Fig. 47, in which it is gradually cooled and where it deposits the remaining portions of the tar.   In order to free

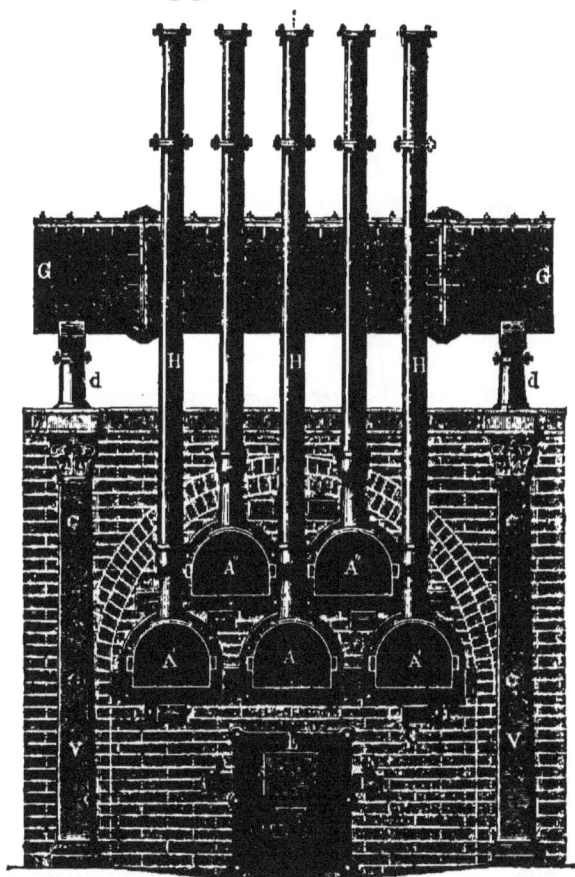

FIG. 45.—Coal-gas Retorts, Front View.   From Roscoe and
Schorlemmer's 'Chemistry.'

it from the last traces of ammonia and the greater portion
of the sulphuretted products derived from the sulphur
in the coal, it is sent through *scrubbers* or wrought iron·
towers (Fig. 48) filled with coke kept constantly moistened
by a fine spray of water; afterwards it passes over layers
of slaked lime and oxide of iron, or mixtures of green

vitriol, sawdust and slaked lime, to remove the carbonic acid
and the last traces of sulphuretted hydrogen, and thence
it flows into the gas-holders. Although the process is
simple enorgh in principle, it nevertheless involves the use

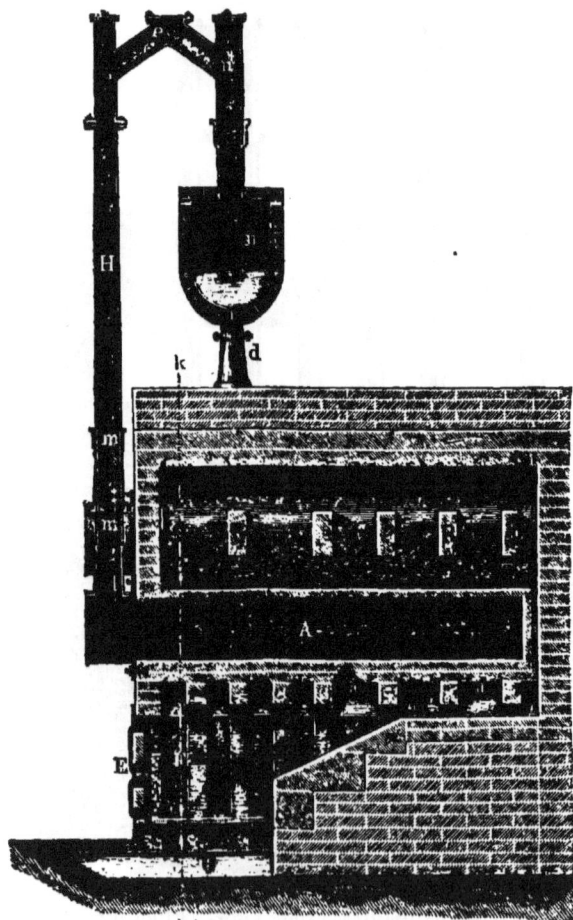

FIG. 46.—Coal-gas Retorts. Longitudinal Section. From Roscoe
and Schorlemmer's ' Chemistry.'

of a considerable amount of machinery in order to bring
about the ready evolution or exsuction of the gas from the
retorts, to facilitate its passage through the scrubbers and
purifiers, and to regulate its distribution through the
mains.

Coal-gas, as delivered to us, is a variable mixture of hydrogen, marsh-gas, and carbon monoxide with certain hydrocarbons. Its general composition may be seen from the following analyses of gas of excellent quality.

	Manchester gas	Gas supplied to Houses of Parliament	Heidelberg gas
Hydrogen	52·71	41·71	41·85
Marsh-gas	31·05	41·88	39·11
Carbon monoxide . . .	4·47	4·98	5·86
Olefines	11·19	8·72	7·95
Nitrogen.	—	2·71	5·01
Carbon dioxide . . .	0·58	—	0·22
	100·00	100·00	100·00

When imperfectly purified it contains small quantities of ammonia and sulphuretted hydrogen, and probably other sulphur compounds, in addition to more or less carbon dioxide. The carbon dioxide diminishes the illuminating power, while the ammonia and sulphur compounds on combustion form acid products which vitiate the air, and rapidly injure the bindings of books and corrode the metal fittings of rooms.

The illuminating power of coal-gas entirely depends upon the hydrocarbons other than marsh-gas which it contains; these are classed together as olefines in the foregoing analyses. One of the most important of these hydrocarbons is known as *ethylene* or *ethene* ($C_2 H_4$); it burns with a highly bituminous flame, the light-giving power of which is due to the ease with which a portion of its carbon is separated in the free state. Other members of the same series of hydrocarbons, notably $C_3 H_6$ and $C_4 H_8$, are probably also present, together with *acetelene* or *ethine* $C_2 H_2$, and minute quantities of *benzene* $C_6 H_6$, all of which contribute to the illuminating value of the gas.

The various inflammable constituents of coal-gas burn

at very different rates. All combustible mixtures of gases
must possess a certain temperature depending on the
nature of the gases, before combination takes place, and
the combination thus set up requires a certain definite in-

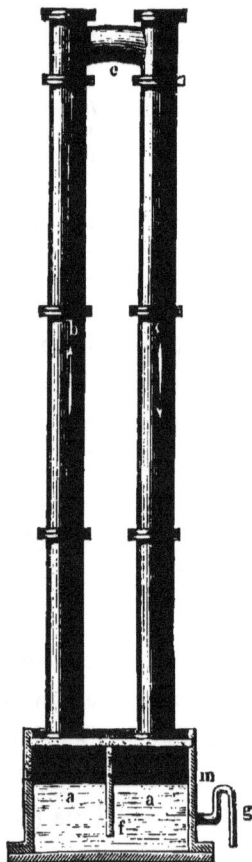

FIG. 47.—Refrigerators or Condensers. From Roscoe and
Schorlemmer's 'Chemistry.'

terval of time to accomplish, also varying with the nature
of the gases. A comparatively low temperature will initiate the union of a mixture of oxygen and hydrogen of the
maximum explosive power, that is, of a mixture which is
entirely converted to water on burning, and the flame, the
visible sign of combination, travels through this mixture

at the rate of about thirty-four metres or thirty-seven yards per second. A mixture of carbon monoxide and oxygen requires a still higher temperature to set up chemical combination under ordinary conditions, and the

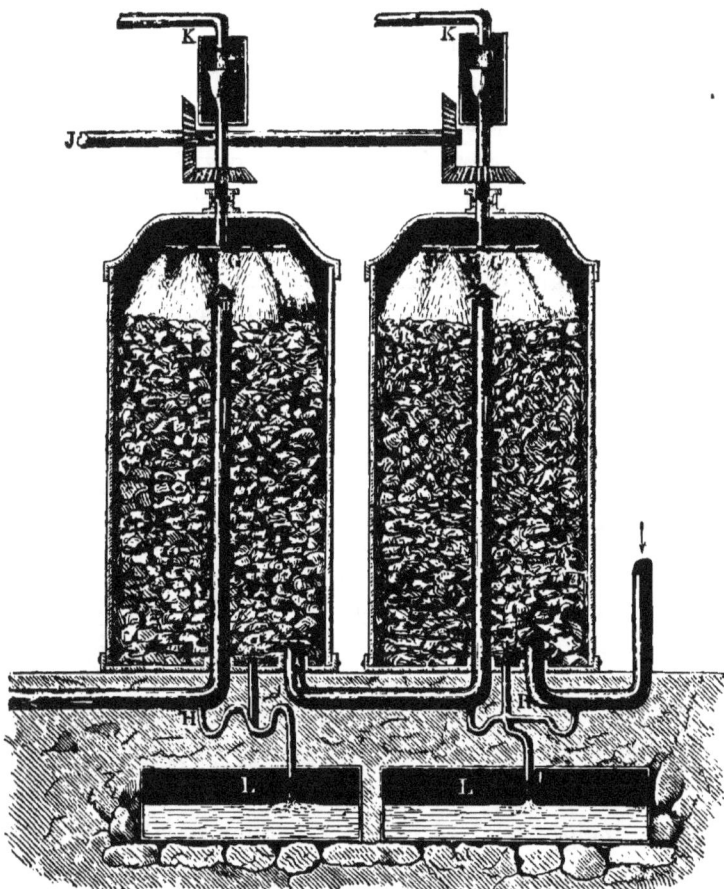

FIG. 48.—' Scrubbers ' for the purification of Coal-gas.   From Roscoe and Schorlemmer's ' Chemistry.'

rapidity of inflammation is much less, only about one metre per second. In the case of a mixture of 1 volume of marsh-gas with $8\frac{1}{4}$ vols. of air, the mixture of maximum explosive power, the rate of inflammation is about half a yard per second, and a comparatively high temperature is needed to initiate the combustion. Hence of the various inflammable constituents in coal-gas, the hydrogen burns most quickly, on account of the comparatively low temperature needed to set up its union with oxygen and the great rapidity with which that union is effected. On the other hand, the hydrocarbons and the carbon monoxide burn most rapidly in the higher parts of the flame.

By sucking out the gases from the different parts of the flame of a gaseous mixture, it is possible to get some idea of the manner in which the constituents comport themselves on burning together. Professor Landolt has studied the flame of Heidelberg coal-gas, the mean composition of which is given on p. 207, with the following results:

Composition of Coal-gas Flame.

Height from Burner in inches	0	0·39	0·79	1·18	1·58	1·97
Total vol. of Air and Gas before burning	127·08	145·43	272·76	327·73	435·30	481·66
Total vol. of Gas after burning	111·41	120·09	245·96	311·37	422·59	461·23
Composition of Flame Gas—						
Hydrogen . . .	22·66	14·95	5·49	15·54	14·50	11·95
Marsh-gas	33·77	30·20	28·34	21·55	11·92	3·64
Carbon monoxide . .	7·34	14·07	14·05	14·58	22·24	25·14
Olefines	7·29	7·49	7·87	7·94	7·05	5·45
Oxygen	0·66	0·78	0·47	—	—	—
Nitrogen . . .	29·41	38·66	140·78	184·23	270·45	307·10
Carbon dioxide . .	1·94	2·34	10·11	14·98	23·76	32·34
Water . . .	8·34	11·60	38·85	52·55	72·67	75·61

At a height of 1·18 inches a sudden break in the continuity of the rate of decrease of the hydrogen is observed, due either to the reducing action of carbon on the water vapour, or to the dissociation of that vapour at the high temperature. The marsh-gas only slowly diminishes up to this point, after which its rate of diminution is very rapid; whilst the olefines burn only in the upper part of the flame. The great increase in the amount of carbon monoxide at $1\frac{1}{2}$ inches is probably due to the action of the strongly heated carbon on the carbon dioxide.

The cause of the luminosity of a coal-gas flame has given rise to much discussion, more especially within the last few years. Sir Humphry Davy was of opinion that it was entirely due to the presence of solid incandescent carbon within the flame, and this hypothesis was generally accepted until a number of observations by Dr. Frankland rendered it not improbable that other causes were concerned in the phenomenon. Recent researches, more particularly those of Dr. Heumann, have, however, rendered it almost certain that Davy's theory is in the main correct, and that solid, incandescent carbon is a *vera causa* of the luminosity of coal-gas flames, and of hydrocarbon flames generally. Soot is practically pure carbon, and the manner in which it is deposited on a cold surface held within the flame is a proof that it has existed therein in the solid state. Moreover soot is non-volatile even at the highest temperatures : hence it cannot exist as a vapour within the flame. Observation also shows that the luminous part of a flame is not absolutely transparent, and hence that this part will cast a distinct shadow in sunlight corresponding to the area containing solid matter. Moreover, by introducing non-volatile matter into a flame of hydrogen or marsh-gas which have no illuminating power, such flames become highly luminous. There can, in short, be little doubt that the presence of solid, incandescent carbon in the coal-gas flame is the main cause of luminosity, although other causes, such as temperature and the comparative density of some of the gases and of their products, may have a subordinate effect.

The by-products obtained in the manufacture of coal-gas, the tar and the ammonia-water, are now-a-days scarcely less important than the coal-gas itself. The ammonia-water furnishes large quantities of ammoniacal salts to be used, among other applications, as food for plants. We thus restore to the vegetation of to-day the nitrogen which existed in the plants of primeval times.

The tar, black and noisome though it be, is a marvellous product, by reason of the scores of beautiful substances which are concealed within it. The distillation of coal and similar substances for the sake of the tar has been carried on, intermittently, for centuries past. So far back as 1694 a patent was granted to one Martin Ecle for ' a way to extract and make great quantities of pitch tar and oyle out of a sort of stone.' In 1742 a second patent was obtained by M. and T. Betton for obtaining ' an oyle extracted from a flinty rock for the cure of rheumatick and scorbutick and other cases ;' and towards the close of the last century the Earl of Dundonald, the father of brave Lord Cochrane, set up tar-ovens on an extensive scale in Ayrshire.

Coal-tar when distilled yields three main products :— *Naphtha, dead-oil*, and *pitch* or *asphalt.* The naphtha on redistillation yields benzene, from which are prepared some of our most beautiful dyes ; the *dead-oil*, as the less volatile portion is termed, furnishes *carbolic acid* or *phenol*, used as a disinfectant and antiseptic, together with *anthracene* and *naphthalin*, all three substances the starting points of new series of colouring matters.

In 1825 Faraday discovered *benzene* among the liquid products obtained by compressing oil-gas, and thereby laid the foundation of this great coal-tar industry. In 1834 the German chemist Mitscherlich found that by the action of nitric acid, the *aqua fortis* of the shops, this benzene was converted into a heavy, oily liquid known to chemists as *nitrobenzene*, and to perfumers as *essence of Mirbane.* Seven years later the Russian chemist Zinin discovered that this substance might be converted into *aniline*, a body previously obtained by Unverdorben from indigo, whence the name, from *anil*, a Portuguese name for the indigo plant. It was long known to chemists that in this aniline an extraordinary colorific power lay latent. Runge in 1835 found that a solution of aniline or of its salts acquired a violet-blue colour with chloride of lime, a reaction still employed by the chemist as a means of detecting the presence of aniline. By dissolving aniline in oil of vitriol, and adding the beautiful orange-red salt bichromate of potash to the solution, a blue-black powder is obtained, out of which Perkin in 1856 isolated the first coal-tar colour *aniline purple*, or 'mauve,' as it was called from the similarity of the colour to that of the flower of the marsh-mallow (French *mauve*). The discovery of this colouring matter marks an era in the history of chemical science. It exercised an extraordinary influence on the development of organic chemistry and on the progress of the chemical arts; theoretical and applied chemistry were knit together in closer union than ever, and dye followed dye in quick succession. After 'mauve' came 'magenta,' and in close attendance followed a brilliant train of reds, yellows, oranges, greens, blues, and violets, all the simple colours of the rainbow in fact.

The colorific intensity of some of these dyes is extraordinary; one ten-millionth of a grain of 'magenta' or 'fuchsin' gives a perceptible colour to a drop of water. This fact teaches us how inconceivably small must be the absolute weight of the chemical atoms. Fuchsin or rosaniline hydrochloride, $C_{20} H_{19} N_3$, HCl, is a compound of high molecular weight; it is $337\frac{1}{2}$ times heavier than hydrogen; hence the absolute weight of a hydrogen atom cannot exceed ·000,000,000,9 of a grain.

It would be impossible to tell how this wealth of

colour is extracted from coal tar without a wearisome
iteration of technicalities, but bare justice to the subject
demands that we should give the story of one at least of
these colours, inasmuch as its discovery constitutes one of
the greatest chemical triumphs of the century.

Most people, doubtless, have heard of Turkey red and of
madder colours, and they know what good, honest, lasting
colours they are. The madder plant, the *Rubia tinctorum*
of botanists, has been cultivated from time immemorial
for the sake of the dye furnished by its roots. Pliny tells
tells us that in his time it was well known 'to the sordid
and avaricious; and this because of the large profits
obtained from it owing to its employment in dyeing wool
and leather.' Originally it was almost exclusively grown
in the neighbourhood of Smyrna and what was European
Turkey. The Moors introduced it into Spain, whence it
found its way into the Netherlands. Alsace and the
district of Avignon have long been celebrated for their
madder; an Armenian named Althen has the credit of
having introduced it into the latter place, and the cruel
story is told that the good citizens of Avignon put up
a statue to his memory to mark their sense of gratitude
for the benefit done to them, on the very day that his
only daughter died in a poor-house. The chief supplies of
madder are obtained from Turkey, Algeria, Italy, France,
and the Netherlands; a few years back it was imported
into this country to the value of 1,250,000*l*. sterling;
and the South Lancashire district alone consumed upwards of 150 tons weekly. It is grown with some little
difficulty; attempts have been made to raise it in England, but with very indifferent success: even under
favourable conditions the yield is not very prolific, and
an acre of land on the average produces only about a ton
of the dried roots.

The nature of the colouring matter of madder has long attracted the attention of chemists, but it is only during the last half century that any very definite knowledge has been gained concerning it. The true colouring agent was isolated in 1827 by two French chemists, Colin and Robiquet, who named it *Alizarin* from *Alizari*, the Levantine name for madder. It is, however, to Dr. Schunck of Manchester that we are principally indebted for our information of the nature of the principles contained in madder. Schunck found that the fresh plant contained no pre-formed alizarin; on slicing the carrot-like root a yellowish liquid can be squeezed out, which on standing becomes red, with the gradual formation of alizarin. The cause of this change is thus explained by Schunck: in the living plant the alizarin exists in combination with a kind of sugar, together with other substances, constituting a body termed *rubian*; under the influence of a nitrogenous ferment also existing in madder, and known as *Erithrozym*, the rubian is split up into its proximate constituents, and thus yields the free alizarin. The same change can be more rapidly brought about by treating the roots with dilute acids. These facts agree with the experience of dyers that madder is far richer in tinctorial matter when kept for some time, and also that its value is increased by treatment with sulphuric acid.

Now that this alizarin might be made artificially was long a dream of the chemists, and that the dream would become a reality was predicted a quarter of a century ago. How this prediction has been fulfilled we proceed to show.

Nearly a hundred years ago an apothecary named Hofmann discovered an acid in Peruvian bark which he called *Quinic Acid*: this acid is also met with in the leaves of ivy, oak, ash, and of many other plants. In investigating the nature of this substance Woskresensky, a Russian chemist working in Liebig's laboratory, found that, on heating, it yielded golden-yellow shining crystals of a body now known to chemists as *Quinone*. The true chemical affinities of this compound remained unknown until 1868, when it was investigated by Dr. Graebe, who pointed out that it stood in a very simple relation to the well-known hydrocarbon benzene; it contains the same number of atoms as benzene, the sole difference being that two atoms of oxygen replace two of hydrogen.

$$C_6H_6 \qquad\qquad C_6H_4O_2$$
Benzene   Quinone

Graebe also showed that chemists were already in possession of bodies which stood to others of the coal-tar hydrocarbons in the same relation that quinone stood to benzene, and he came to the conclusion that alizarin was really a derivative of one of these quinones. It remained to ascertain what was the parent hydrocarbon. Graebe and his collaborateur Liebermann in the first place heated a quantity of alizarin with finely powdered zinc, and obtained a hydrocarbon crystallising in fine pearly plates, with a beautiful blue fluorescence; this they recognised as identical with a body long known to chemists as having been discovered in 1832 by Dumas and Laurent, and called *Anthracene* from its occurrence in coal-tar (ἄνθραξ, coal). Now would this anthracene yield a quinone? There was no need for Graebe and Liebermann to try the experiment; it had already been done for them by Anderson, who in 1861 had prepared the compound from anthracene by boiling it with nitric acid and had named it *Oxanthracene* since it differed from anthracene by containing two atoms of oxygen in place of an equal number of hydrogen atoms, standing in fact to the hydrocarbon in precisely the same relation that quinone stands to benzene :

$$C_{14}H_{10} \qquad\qquad C_{14}H_8O_2$$
Anthracene   Oxanthracene or Anthraquinone

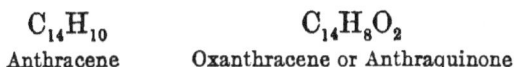

Hence Graebe gave it the systematic name of *Anthraquinone.*

Now alizarin differs from anthraquinone by containing two additional atoms of oxygen :

$$C_{14}H_8O_2 \qquad\qquad C_{14}H_8O_4$$
Anthraquinone   Alizarin

There was no difficulty in effecting the addition of the oxygen. In the first place, by the action of bromine on anthraquinone, two atoms of bromine were put in the place of two of hydrogen :

$$C_{14}H_8O_2 \;+\; 2Br_2 \;=\; C_{14}H_6Br_2O_2 \;+\; 2HBr$$
Anthraquinone   Bromine   Bibromanthraquinone   Hydrobromic acid

Now to get the oxygen in and the bromine out—this could be done by boiling with potash :

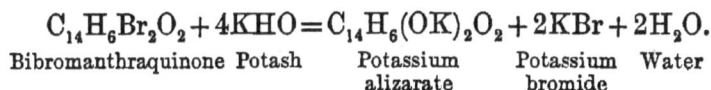

$$C_{14}H_6Br_2O_2 + 4KHO = C_{14}H_6(OK)_2O_2 + 2KBr + 2H_2O.$$
Bibromanthraquinone   Potash   Potassium   Potassium   Water
                              alizarate   bromide

Two atoms of potassium (K) were now to be replaced by two of hydrogen, and alizarin is formed. This was readily effected by treating the potassium alizarate with sulphuric acid :

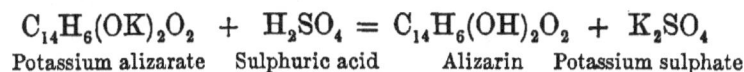

$$C_{14}H_6(OK)_2O_2 \;+\; H_2SO_4 \;=\; C_{14}H_6(OH)_2O_2 \;+\; K_2SO_4$$
Potassium alizarate   Sulphuric acid   Alizarin   Potassium sulphate

In this manner, then, Graebe and Liebermann accomplished the artificial formation of this important colouring matter; a discovery of which it is scarcely possible to exaggerate the significance. In any case it would have been interesting as constituting the first instance of the artificial production of a vegetable dye, but when the body so formed happened to be the very chief of vegetable dyes, ·

the interest became extraordinary. Scientifically speaking the process was complete, but to produce alizarin on a manufacturing scale much still remained to be done. A dozen chemists immediately attacked the problem, and again Mr. Perkin was in the van, with Messrs. Graebe, Liebermann, and Caro. The modification proposed by these chemists consisted simply in the substitution of the cheap oil of vitriol for the expensive bromine. By heating the anthraquinone, obtained from the anthracene of coal-tar, with sulphuric acid, neutralising the free acid with lime, converting the sulpho-acid into a soda salt, heating it with soda to a high temperature and decomposing with an acid, alizarin is now obtained by the ton in the form of a yellowish mud equal in dyeing power to eight times its weight of the best madder. The effect on the value of madder has been enormous; it is rapidly dropping to a price at which it can no longer be produced, and hundreds of acres are ready to receive some new crop. We have been told that he who makes two blades of grass to grow where only one blade grew before, is a benefactor to his kind; how great then is his service who makes the remnant of a primeval vegetation do the duty of the vegetation of to-day?

But the theoretical chemist has achieved a still greater triumph in that he has shown how this beautiful colouring matter can actually be built up from the black carbon and the colourless invisible hydrogen. At a very high temperature carbon can be made to combine with hydrogen to form the hydrocarbon *acetylene*, from which, by heating, simple benzene can be obtained. By heating together benzene and ethylene, anthracene itself can be formed, from which we get the alizarin in the manner just described.

1. Acetylene by direct union of carbon and hydrogen. (Berthelot, 1862.)

$$C_2 + H_2 = C_2H_2.$$

2. Benzene from acetylene by heat. (Berthelot, 1866.)

$$3C_2H_2 = C_6H_6.$$

3. Anthracene from benzene and ethylene. (Berthelot, 1866.)

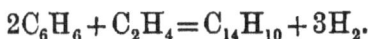

$$2C_6H_6 + C_2H_4 = C_{14}H_{10} + 3H_2.$$

4. Anthraquinone from anthracene. (Anderson, 1861.)

$$C_{14}H_{10} + O_3 = C_{14}H_6(OH)_2 + H_2O.$$

5. Anthraquinonsulphonic acid from anthraquinone.

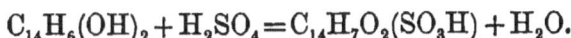

$$C_{14}H_6(OH)_2 + H_2SO_4 = C_{14}H_7O_2(SO_3H) + H_2O.$$

6. Conversion of the acid into soda salt.

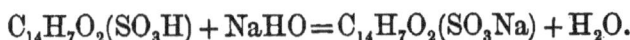

$$C_{14}H_7O_2(SO_3H) + NaHO = C_{14}H_7O_2(SO_3Na) + H_2O.$$

7. Formation of monoxyanthraquinone.

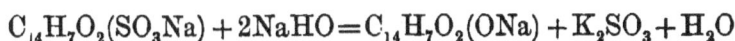

$$C_{14}H_7O_2(SO_3Na) + 2NaHO = C_{14}H_7O_2(ONa) + K_2SO_3 + H_2O$$

8. Conversion into soda alizarate.

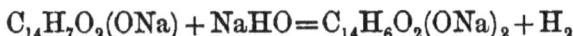

$$C_{14}H_7O_2(ONa) + NaHO = C_{14}H_6O_2(ONa)_2 + H_2$$

9. Decomposition of soda salt by sulphuric acid

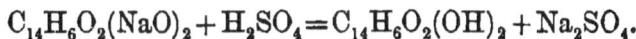

$$C_{14}H_6O_2(NaO)_2 + H_2SO_4 = C_{14}H_6O_2(OH)_2 + Na_2SO_4.$$

Alizarin

But there is still another story of coal-tar to be told. Among the many curious substances which that wonderful fluid contains is a beautifully white, wax-like body called *paraffin*, a hydrocarbon discovered in 1830 in the tar obtained by distilling beech-wood, by the German chemist Reichenbach, who described its main properties and gave it its name in allusion to its remarkable stability and chemical indifference (*parum affinis*). Two manufacturing chemists, Butler in 1833, and Du Buisson in 1845, worked patents for its extraction from coal-tar, but with little or no commercial success, and what we now know as the 'Paraffin Industry' mainly owes its origin to the genius and energy of Mr. James Young. As early as 1848 Mr. Young had worked a small petroleum spring occurring in a coal mine at Alfreton in Derbyshire, and had produced oils suitable for burning and lubricating purposes, but the spring 'gave out' as the Americans say, by which they imply that it ceased to give out anything at all, and Mr. Young sought to obtain these oils artificially by distilling coal. After many trials with numerous materials, he lighted on the now famous Boghead Cannel, and set up his works, in conjunction with Mr. Meldrum and Mr. Binney the geologist, at Bathgate in Scotland, in the very centre of the Torbane Hill coal-field. The present magnitude of this industry is something extraordinary; indeed, its rate of growth is without parallel in the history of British manufactures. In Scotland alone there are some sixty paraffin oil works. One of these, the Addiewell works, a younger but a bigger sister of the Bathgate works, occupies a site of nearly forty acres, and the land leased in connection with it extends to over 3,500 acres. Here 3,000 tons of shale are weekly distilled from upwards of 400 retorts, yielding nearly 120,000 gallons of crude oil. Among the various works in Scotland about 800,000 tons of shale are distilled per annum, producing nearly 30,000,000 gallons of crude oil, from which about 12,000,000 gallons of refined burning oil are obtained, in addition to large quantities of naphtha, solid paraffin, ammonia, and other chemical products.

The operation of distilling shale is very similar to that already described in connection with the manufacture of coal gas. The shale is first crushed between rollers, and is thrown into cast-iron retorts and distilled at a heat scarcely exceeding within the retort the melting point of plumbers' solder. At this comparatively low temperature a large proportion of the volatilised product is condensed into oils, although much resists liquefaction and is collected in large gas-holders to be used for lighting the works, or as a source of heat. The dark brown crude oil is separated from the ammonia water and is redistilled, agitated first with oil of vitriol and then with soda and again distilled. That which first passes over is known as 'petroleum spirit' or 'benzoline,' and is largely used in the arts as a substitute for turpentine; the next fraction is the burning oil, whilst the third is known as 'heavy oil,' and is separated into solid paraffin or paraffin wax and lubricating oil. To effect the separation, the 'heavy oil' is first strongly cooled by a very beautiful and ingenious apparatus identical in principle with that by which M. Pictet has succeeded in liquefying the air and hydrogen. The paraffin crystallises out and is subjected to pressure to squeeze out as much as possible of the dark coloured oil; the semi-solid mass is then washed with petroleum spirit which dissolves out the last traces of the oil, after which it is again pressed and, if required perfectly white, is melted with oil of vitriol. A very old metallurgical operation known as 'liquation,' in which advantage is taken of the different fusibilities of metals to effect their separation, has recently been extended to the purification of paraffin. The impure scales are pressed in a cake, placed on sloping shelves, and heated nearly to the softening point. The more readily liquefiable portion gradually melts out and drains away, carrying with it the colouring matter and leaving the solid paraffin nearly pure and white. This method is said to be due to the intelligence of a workman, who noticing how a small mass of scales which had fallen over a steam pipe gradually became whiter and harder by the melting out of a portion, was induced to make experiments which demonstrated the practicability of the process on the large scale.

Five-and-twenty years ago scarcely a dozen persons had seen this paraffin, and now it is turned out by the ton to be fashioned into candles, some of which are delicately tinted with colours also obtained from coal-tar. Years ago the great chemist Liebig with wonderful pre-vision said that it would certainly be esteemed one of the greatest discoveries of the age, if any one could succeed in condensing coal-gas into a white, dry, solid, odourless sub-stance, portable, and capable of being placed on a candle-stick and burned in a lamp. And in this solid paraffin which, chemically speaking, is the omega of a series of bodies of which marsh-gas is the alpha, we see the ac-complishment of this great discovery.

At the present time the paraffin-oil manufacturers of Scotland are hard pressed by their American rivals, who, in the petroleum springs of Pennsylvania have an abundant supply of oil, easily worked and more readily refined than the artificial product; it is only by skilful management and the constant and intelligent application of science to the improvement of processes, and to the utilisation of what would be otherwise wasted that the home manufacturers are able to hold their own. Paraffin can be got from peat; and many years ago, during the time of the Potato Famine, an honourable member astonished the House by displaying a white, wax-like candle which he affirmed had been made from the material of an Irish bog. Ireland's peat bogs have been matter for derision to the Saxon for centuries, and behold they were to become the means of her regener-ation! Her 4,000,000,000 tons of dry peat would furnish

5,000,000 tons of paraffin, as much ammonia, and as many
gallons of oil. So calculated her sanguine friends ; but
alas ! experience has shown that Ireland's means of grace,
if it is to be found in her bogs, is still in the dim future,
for so long as the stream from the American wells continues, her bogs are much more likely to swallow up trea-
sure than to yield it.   Signs are not wanting that, however, the tide will turn.   In the meantime, it is some
consolation to know that peat, like port-wine, improves
with age.   Let us hope then that Fortune, who has
certainly looked rather askance at our poor sister in the
past, will ultimately grant her, if not the overflowing
wealth which springs from vast supplies of mineral fuel,
at least the possibility of an extensive and remunerative
industry.

CHAPTER VII.

COAL AS A SOURCE OF WARMTH.

COAL, when employed in the production of heat is popularly regarded as possessing the two-fold character of a source of warmth and a source of power. As the first, it is employed for domestic purposes, and in many manufacturing operations in which high temperatures are needed ; as the second, it is used in various machines and notably in the steam engine, for the production of mechanical work. The scientific accuracy of this division may be questioned. No one doubts that coal· is a source of power when it enables a steam hammer to weld together two bars of iron, by forcing their particles, in spite of a strenuous resistance, into such close proximity that they cohere. The coal burnt in the furnace in which the iron was smelted, is, on the other hand, ordinarily regarded as used only as a source of warmth, yet it too is really employed to overcome a resistance. The iron ore upon which it acts is decomposed, and the particles of iron are forced away from the other substances to which they were united by the bonds of chemical union. The difference between the two cases lies, not so much in the nature of the task accomplished, as in the fact that the heat which is directly applied when the iron is to be smelted, acts, when it is welded, indirectly through the agency of an elaborate machine.

The division of our subject which the view above re-

ferred to suggests, is, however, in many respects convenient, and we shall perhaps be justified in accepting it,
if we regard it as historical rather than scientific. The
value of coal as a fuel was keenly appreciated long before
any conscious attempt was made to employ it as a source
of power.

Regarding, then, mineral fuel in the first place as a
source of warmth, I shall commence by detailing some of
the principal facts which are known with respect to its
early employment in this capacity, tracing the steps by
which it has to so great an extent superseded wood, charcoal, peat, and other substances employed for the same
purposes. Having seen it firmly established and in
general use as a fuel, we shall next discuss the scientific
principles on which an estimate of its value can be
founded. The results to which we are led will serve as
bases for the second branch of our enquiry, in which we
shall investigate the relations between the heat produced
by coal and the amount of work which that heat might
be, and actually is, made to perform.

The ancients appear to have been acquainted with
coal, and with the fact that it would burn, but they do
not seem to have turned their knowledge to much practical account. Theophrastus, a Greek writer who flourished
about two hundred years before the Christian era, in a work
entitled 'The Book on Stones,' describes an earthy substance which would kindle and burn, and which was used
by smiths. There can be no doubt that he refers to coal,
and that this is the earliest passage in which that substance is expressly mentioned.

The word *coal* is indeed frequently used in our trans-
lation of parts of the Bible which were doubtless written
long before the time of Theophrastus; but this is explained, not by the assumption that the authors were ac-

Q

quainted with mineral fuel, but by the fact that the translators used the word coal in a much wider sense than that to which it is at present generally confined.

Coal, or cole, originally meant any substance which was used as fuel. Thus Sir John Pettus in his '*Fodinæ Regales*,' published in 1660, mentions two kinds of ' coale : '

' 1. Black, such as is burnt or charkt.

' 2. White, which is only baked in an oven to make it dry for fewele.'

The two substances here mentioned are evidently charcoal, and white or uncharred wood, so that with this author ' coale ' meant wood fuel. Indeed, in earlier times, when the substance to which alóne we give the name of coal was referred to, it was generally called earth cole, or pit cole, or (since it was brought to London in large quantities from Newcastle by sea) sea cole. It was only when the use of this earth coal became so general and its importance so great, as to make it rank as the first of fuels, that these prefixes were dropped, and the word coal was applied to mineral fuel alone.

Several supposed references to coal have been noted in classical authors of a date later than Theophrastus, but in the case of some of these it is doubtful whether coal or another substance is meant. From other passages of a less doubtful character we gather that blocks of coal were at times used as landmarks.

That the early inhabitants of Britain were acquainted with coal is proved by what Prof. Hull calls 'incontestable evidence,' viz., the discovery in Monmouthshire of a flint axe stuck in a vein of coal, which is there exposed on the surface of the ground, and which would therefore be easily discovered and worked. Axes or picks of solid oak have been found in some old excavations near Stanley in Derbyshire, and stone hammer heads, wedges

of flint, and wooden wheels, in coal workings, which are
evidently of great antiquity, near Ashby-de-la-Zouch.

Cæsar makes no mention of coal in his description of
Britain, but the Romans appear to have employed it at
a later date. Heaps of cinders mingled with Roman remains have been found at Castlefield near Manchester,
at North Brierly in the West Riding, near Newcastle,
and in other places. Near Wigan in Lancashire an ancient coal mine was some years ago discovered, in which
the seam had been excavated in a manner ' different from
anything within the experience of the miners of the district.' [1] These workings, which local tradition ascribed
to the Danes, Prof. Hull is inclined to attribute to
the Romans. We have, however, no documentary evidence of the employment of English coal earlier than
853 A.D. It is first mentioned in a grant of lands made
by the Abbey of Peterborough, a record of which is preserved in a Saxon chronicle. The word employed—
' græfan '—is used in no other place in Saxon literature.
It is supposed to be connected with the German *grab*, a
trench, and with our own word *grave*, which means a trench
used for interment. It thus probably referred to the method by which coal is obtained.

Coal is twice mentioned in the Bolden Book, a census
of part of the northern counties, published in the reign
of Henry II. In 1130 the Bishop of Durham made a
grant of lands to a collier to provide coals for a smith.
In 1239 the King granted licence to the good men of
Newcastle to dig for coals in the neighbourhood of their
town, and we are told that in 1281 Newcastle had a
considerable trade in coal.

Mineral fuel did not, however, come into general use
without meeting with considerable opposition. In 1306

[1] *Coal Fields of Great Britain*, p. 16.   E. Hull, F.R.S., 1873.

Parliament petitioned King Edward I. to prohibit the
use of coal in London. The citizens were so offended at
the 'sulferous smoke and savor of the firing' that the
King complied with their request, and the penalty for the
use of coal was the destruction of the furnaces or kilns in
which it had been used. We are also told that many years
afterwards 'the nice dames of London would not come
into any house or room where sea-coals were burned, nor
willingly eat of the meat that was either sod or roasted
with sea-coal fire;'[1] and even so late as 1679 we learn
that Mr. Locke, finding himself affected with an asthma,
'could not tarry long at London, the sea-coal that is
burnt there being so very offensive to him.'[2]   Such
prejudices were rife in France up to a much later date.
In 1714 the Parisians refused to allow coal to be used in
Paris. In 1769—little more than a hundred years ago—
when wood was very dear in Paris, 'some English mer-
chants conceived the idea of sending cargoes of coal from
the English Collieries to supply the scarcity of fuel in
France. Vessels were dispatched with coal from Newcastle, which was sent up the Seine and soon reached
Paris. The coal which was sent from England was at
first used by the common people, and then by some of the
higher classes in the stoves and fire-places of ante-chambers.
There was a general outcry. The unfortunate mineral
fuel was accused of vitiating the air, of soiling the linen
when drying, of causing affections of the chest, and, worst
of all, impairing the delicacy of the female complexion.
Complaints being unceasing, the Academies of Science
and Medicine were appealed to,'[3] but though they declared
in favour of the British fuel, Parisian opinion remained

[1] Stow's *Annals*: by Homes, p. 1502. London, 1652.
[2] Report of Commissioners on Coal, p. 8.
[3] *Underground Life, or Mines and Miners*, p. 3. Louis Simonin. Trans-
lated by H. W. Bristow, F.R.S. Chapman and Hall, 1869.

unaltered, and the obnoxious mineral was once more expelled.

In spite, however, of the edicts of kings and the prejudices of the people, the employment of coal as a fuel became perforce more and more general. Thus though King Edward I. forbade its use in London, in 13C6, it was used at the coronation of his successor only fifteen years later. In 1367 coals were consigned to the Clerk of the Works at Windsor Castle, and in 1379 the first government tax was laid on coal. The Parisians must have been more sensitive than the rest of their countrymen, as coal was exported from England to France as early as 1325; and in the reign of Edward VI. an English writer speaks of 'that thinge that France can live no more without, than the fyshe without water; that is to say, Newe-castele coles, which without that they can nother make stele worke, nor metall worke, nor wyer worke, nor goldsmythe worke, nor gonnes, nor no manner of thinge that passeth the fier.' [1] In England the supporters of coal had a century later gained sufficient strength to become in their turn abusive. During the war between Charles I. and the Parliament, as Newcastle was strongly Parliamentarian, the King prohibited all trade between that town and London. A coal famine was produced in London by this measure, during the distress consequent on which a pamphlet was published in the form of a discourse between Sea-coale, Char-coale, and Small-coale, three persons engaged in different branches of the fuel trade. In the course of their discussion Char-coale remarks, 'Were Sea-coale's whole lineage burnt out, I and my kindred could supply the defect of the city far more cleanly and sweetly than this vaporous sooty-faced Sea-coale can doe possibly.' To whom Sea-coale replies, 'It is impossible, but thy

[1] Report of Commissioners on Coal, iii. 4.

malice must break out against me, thou spark of iniquity,
thou suffocating bastard, begot between a Sussex Ironmon-
ger and a Kentish long-tail.   Thou serve the city as well
as Sea-coale ?   Thou art burned out like a farthing candle
immediately.   One bushell of mee being worth a loade of
thee, or a hundred of your cousin Billets and Faggots
that you so boast of.'[1]

The necessity which compelled the general use of coal
in spite of real or fanciful objections, was the ever increasing scarcity of wood.   Year by year the dense forests,
which in the time of the Romans covered these islands and
a considerable portion of the continent of Europe, had
been disappearing as the wants of the growing population
brought more and more land into cultivation.   The stores
of wood fuel, which to earlier generations must have seemed
inexhaustible, gave way, and in the reign of Elizabeth
the necessity either of preserving the forests, or of finding
some substitute for them, seems to have been very generally felt.

Thus in 1570 Queen Elizabeth was petitioned that
'the bloomaries,' or furnaces in High Furness might be
abolished, on account of the quantity of wood which was
consumed in them for the use of the mines, to the great
detriment of the cattle.   Harrison writing a few years
later says, ' of colemines we have such plentie in the north
and western parts of our island as may suffice for all the
realme of England, and so must they doo hereafter indeed,
if wood be not better cherished than it is at this present.
And to saie the truth, notwithstanding that verie manie of
them are carried into other countries of the maine, yet their
greatest trade beginneth now to grow from the forge into
the kitchen or hall, as may appeere alreadie in most cities

---

[1] *Sea-coale, Char-coale, and Small-coale*: published 1643.   Reprinted
at Newcastle 1843. See Report of Commissioners on Coal, iii. 7.

or townes that lie about the court, where they have but little other fewell except it be turffe or hassocke.' [1]

It is not improbable that the objections to the use of coal in earlier times were partly caused by the fact that it was obtained near the surface and was of inferior quality. As time went on it was won at greater depths, and greater attention was paid to the art of mining. A treatise on the subject of metal mining was published by Agricola in the middle of the sixteenth century, and John Dee the celebrated astrologer, writing in 1570, claims to be the inventor of a new art, by means of which the depth and position with respect to objects on the surface of an underground working could be determined. This art he says will be useful among others to 'Diggers for cole,' and he relates that he invented it 'at the request of two Gentlemen, who had a certain work (of gain) underground,' and who were 'at controversie under whose ground, as then, the work was.' [2]

The development of the coal trade under the Tudors was indeed very marked, and we find that at that period the Yorkshire Coal Field was important. Thus Leland, who visited Rotherham near Sheffield about 1550, tells us that 'though betwixt Cawood and Rotherham be good plenti of wood, yet the people burne much yerth cole, by cawse hit is plentifully found ther and sold good chepe;' and he adds that 'a mile from Rotherham be veri good pittes of cole.'

Thoresby, the historian of Leeds, states on the authority of a manuscript note by W. Vavasour, of Haslewood, Esquire, that when King Henry VIII. made progress to York in 1548, he was told by Bishop Tonstall that there

[1] Preface to Holinshed's *Chronicles of England, Scotland, and Ireland*, i. 397. London, 1805.

[2] Mathematical Preface of John Dee, prefixed to the first English Edition of Euclid.

were within ten miles of Haslewood Park no less than twenty-five coal mines and three forges.[1] Among the mines those at Temple Newsam, Roundhay Park, Seacroft, and Harwoodmore are mentioned; and among the forges that which had been established long before and which still exists at Kirkstall.

Early in the next century much attention began to be given to the problem of the possibility of employing coal instead of charcoal in the process of iron smelting. The quantity of wood used for this purpose was so great that in the reign of Elizabeth four acts were passed to check its consumption. In a tract written in 1629 entitled ' A Relation of some Abuses committed against the Commonwealth,' an ironmaster in the neighbourhood of Durham is accused of having ' brought to the ground (to omit all underwood) above 30,000 oaks in his lifetime; and' adds the author ' (if he live long enough) it is doubted that he will leave so much timber in the whole country as will repair one of our churches if it should fall.'

Under circumstances such as these, it was evident that he who could successfully introduce the use of sea-coal into a manufacture which was thus devastating the forests, would not only confer a great benefit upon the country, but would himself reap a rich reward. Golden dreams led one and another to make the attempt. One of the earliest of these speculators, Simon Sturtevant, announced that he had discovered a method by means of which the smelting of as much iron as had hitherto required an outlay of 500*l.* in charcoal could be accomplished with from 30*l.* to 50*l.* worth of sea-coal. His

---

[1] Fuller's *Worthies,* from a MS. at Haslewood. Quoted by Thoresby, *Ducatus Leodensis,* Preface, 1714. Second Edition with notes by Whitaker, Leeds, 1816.

method, however, does not seem to have realised his expectations. The difficulties to be overcome were very great. Iron had previously been smelted in small quantities, and to substitute coal for charcoal successfully it would have been necessary to employ larger furnaces and more powerful bellows than were then in use. The sulphur, too, of which coal contains a small proportion, affected injuriously the quality of the iron produced. The device now employed of getting rid of the sulphur by converting the coal into coke was not at that time resorted to.

Lord Dudley, commonly called Dud Dudley, who when a lad of twenty left Balliol College, Oxford, to superintend his father's ironworks, devoted his life to an attempt to overcome these obstacles. He had to contend not merely with the difficulties inherent in his task, but also with the opposition of the charcoal ironmasters, by whom he was driven from a furnace with which he claimed to have attained some success. Undeterred by this disaster, he built a larger furnace at Sedgley in Staffordshire, and is said to have produced iron at from 20 to 40 per cent. below the ordinary market price. Misfortune however, dogged his steps. A Bristol firm in which he was a partner failed, his most enterprising colleague emigrated to Ireland, and finally a renewal of his patent was refused. In short as the result of forty years of toil, difficulty, and danger, he does not seem to have done much to further the object which he had at heart.

Three quarters of a century later (1735) Dudley's dreams were realised, and Abraham Darby applied coke to the smelting of iron with complete success at the celebrated Colebrook Dale Iron Works in Shropshire. The impetus which this event gave to the coal trade was increased by the introduction of the steam engine.

Newcomen invented his engine in 1705, and James Watt commenced his famous series of improvements on the Newcomen engine in 1763. The subjoined table gives the total quantity of coal estimated by the Commissioners on coal to have been produced in the United Kingdom in the years named. It shows the rapid development of the trade during the century in which coal was first successfully used both in the smelting of iron, and the production of power.

TABLE I.

Years	Estimated produce of Kingdom
	Tons
1660	2,148,000
1700	2,612,000
1750	4,773,828
1770	6,205,400
1780	6,424,976
1790	7,618,760
1800	10,080,300

We have now rapidly run over the history of the use of coal from the earliest to comparatively modern times. We have reached a period in which it was not only in general demand for domestic purposes, but was also beginning to be employed in some of those industries, to the very existence of which it is now considered essential. Its value had been proved by experiment conducted on the largest scale, and it took rank by common consent as the most convenient source of heat at our disposal. At this point then we may fitly enter upon another branch of our enquiry, and dismissing for the present the tests of the value of coal afforded by its increased and increasing use, we may proceed to apply those which are based upon the more delicate methods of investigation employed in the laboratory.

. No more important question can be asked with respect

to any fuel than this : What amount of heat will the combustion of a pound of it produce ?

The answer evidently involves a knowledge both of the
general principles on which heat is measured, and of the
application of those principles to the determination of the
heat produced by a burning body.   Let us devote a short
time to each of these subjects in turn.

Alterations in the temperature of bodies may be
detected and measured by various methods.   Our own
sensations of heat and cold often give us some information on this point, but they cannot be trusted for accurate
observation, and we are accustomed to assist them with
instruments.   The most important of these, the mercurial
thermometer, is too well known to require description.

It is however necessary to remember that although
this and other instruments enable us to compare the
alterations in temperature which different bodies may
undergo, these measurements of temperature do not
directly inform us of the quantities of heat which enter or
leave the bodies as they become warmer or colder.

If two similar closed vessels containing the one some
warm water, and the other the same weight of oil at the
same temperature, were placed in the open air on a cold
day, the liquids would cool, and if when their temperatures
had ceased to fall they were once more determined, they
would be found to be again the same.   It would however
be observed that the water would take longer to cool than
the oil, and if by chance the vessels had been placed upon
a block of ice, that containing the water would have
melted the larger hole in the ice.   These circumstances
would lead to the suspicion (which closer investigations
would confirm) that the water had lost the larger quantity
of heat, though each liquid had been cooled through the
same number of degrees of temperature.   The following is

an experiment by means of which the fact that equal
weights of different substances, when cooled through the
same range of temperature, do not give out equal quantities of heat, may be rendered evident to a large
number of persons at once.

A small glass cell containing some water and a thermometer is placed in front of a lime-light lantern, and by
means of a lens an enlarged image of the cell and its

FIG. 49.—Lantern Calorimeter.

contents is formed upon a screen. The thermometer is so
constructed that the whole of it may be visible. The stem
is bent round parallel to the bulb, it is graduated for 20°
only, from 60° to 80° F., and the graduations are engraved
upon a piece of glass to which the instrument is attached,
and by means of which it is supported. A large glass
trough filled with water is placed between the cell and the
lantern. This allows the light to pass through, but cuts
off so much of the heat that the thermometer remains
stationary though exposed to the full blaze of the lime-
light. Previous to the performance of the experiment the
temperature of the water in the cell must be adjusted to a
little over 60° F. (say 60½°), and equal weights of water and
quicksilver contained in two separate tubes must be heated
by immersion in a vessel of boiling water. If the tube in
which the water is placed be graduated, any loss which

may occur from the evaporation of the water contained in it while heating may be compensated by the addition of a few drops of that in which it is immersed.

Now let the hot mercury be thrown into the little cell, and let the water and mercury be stirred with a glass rod to equalise their temperatures. The thermometer will register only a very small elevation in the temperature of the water ; as soon however as the hot water is poured in it rises briskly. If after the introduction of the mercury the thermometer rises from $60\frac{1}{2}°$ to $61°$, it will when the hot water is added rise to nearly $74\frac{1}{2}°$. In other words, the heat given up by the water produces an increase in the temperature of the cell and its contents twenty-seven times as great as that due to the heat given up by the mercury, though the latter was cooled through a slightly larger number of degrees, say from $212°$ to $61°$ instead of from $212°$ to $74\frac{1}{2}°$. Conversely, the experiment proves that if equal weights of water and mercury are each raised through $1°$ F., a larger quantity of heat will be expended in the case of the water than in that of the mercury. Experiments conducted on the same principle with other substances prove that they differ as much in this as in other respects. Equal amounts of heat produce in them very different increments of temperature ; equal alterations in temperature require very different amounts of heat.

We cannot measure a quantity of heat by merely saying that it is the amount necessary to raise a pound of matter through so many degrees ; we must mention the particular kind of matter to which we allude. The most convenient plan therefore is, once for all, to choose some standard substance to which all measurements of heat may be referred. The substance so selected is water, and the *unit of heat* is defined to be *the quantity of heat necessary to raise one pound of water from* $32°$ *to* $33°$ F.

A pound of mercury would be raised through one degree by one-thirtieth of this unit of heat, a pound of iron by one-ninth, a pound of charcoal by one quarter. It is used in measurements of heat just as the yard is used in measurements of length, or the gallon in those of volume. All three are arbitrary and conventional standards, with which, by general consent, volumes, lengths, or quantities of heat are compared. The French use a different unit of heat just as they use a different unit of length. There is no particular reason why the unit above defined should be selected rather than any other, except that the Fahrenheit thermometer is in general use in this country, and that water is a convenient substance to employ in heat-measuring experiments; but there is also no particular reason why the yard should be the unit of length except that Parliament has so decreed. Both units are valuable only in so far as they are generally used.

Having thus determined on a method of measuring heat, we may now proceed to employ it in estimating the quantity of heat produced by the combustion of a given weight of any substance. It will only be necessary to burn the substance in such a manner that the heat produced may be communicated to a known weight of water. If the experiment is so conducted that no heat is lost while the body is burning, and that the rise in the temperature of the water can be correctly measured, results of great accuracy can be obtained. Different observers have adopted different experimental methods, but it will be sufficient to describe that employed by Prof. Andrews of Belfast.

The apparatus he used is represented in Fig. 50. It consisted of a vessel made of thin copper (a) the capacity of which was about three and a half quarts. This was filled with oxygen and the substance to be burnt was

placed in a small platinum dish (*b*), which was suspended
from the copper cover of the vessel by three platinum
wires. The whole was placed in a larger vessel containing a known weight of water, and fitted with a
cover pierced with two holes, through which passed a
thermometer (*c*) to measure the temperature of the water,
and a rod (*d*) by which the vessel was kept in its place.

FIG. 50.—Prof. Andrews' Apparatus for measuring the heat produced
by combustion.

In order to avoid loss of heat it was necessary to make
arrangements for igniting the substance when the vessel
in which it was burnt was surrounded by water. To accomplish this, advantage was taken of the fact that an electric current heats a wire through which it passes, and *cæte-
ris paribus* heats it the more strongly the thinner it is.

A platinum wire which passed through the cover of
the vessel (*a*), but which was carefully insulated from it,
was connected by a very fine wire with the platinum dish.
The wire, the dish, the wires which supported it, and the
vessel could thus be included in a galvanic circuit.

When the thin wire was heated the combustible

substance would ignite, and when the electric current had done its work it could immediately be stopped. The heat produced in the wire would thus be sufficient to induce combustion, not sufficient to render the result of the experiment inaccurate. When the combustion was complete the vessel was moved up and down by means of the rod (d), the water was thus stirred, and its temperature having been made uniform, was determined by a delicate thermometer.

The water used in each experiment was accurately weighed, and allowance was made for the fact that not only the water but also the metal vessels had been heated. When a gas such as hydrogen was to be burnt an apparatus of somewhat different construction was used.

The following table contains the results of Andrews' experiments on several substances of which we shall have hereafter to speak again :

TABLE II.

Substance	Product of Combustion	No. of units of heat produced by combustion of one pound of the substance
Hydrogen	Water	60,986
Carbon	Carbonic Acid	14,220
Sulphur	Sulphur Dioxide	4,153

Prof. Andrew's experiments were made rather more than thirty years ago, and others have since determined the heat of combustion of various substances. The following table, which gives the heat evolved by the combustion of hydrogen, according to different observers, will give an idea of the degree of accuracy which can be obtained :—

TABLE III.

Observer	Date	No. of units of heat evolved by combustion of 1 lb. of hydrogen
Andrews.	1845	60,986
Favre and Silbermann . . .	1845	61,806
Thomsen	1872	61,636
Than	1877	61,447
Schuller and Wartha . . .	1877	61,916

Assuming then that the heat of combustion of bodies can be correctly determined, we must pass on to the application of the knowledge so acquired to the special case of coal.

It would of course be possible to perform with coal experiments similar to those which have just been described, but the practical difficulties involved in such an investigation have led to the adoption of various less direct methods. One of these, which has been very generally followed in determining the heat of combustion of coal, has been to analyse it, i.e. to find out what proportion of carbon, hydrogen &c., it contains, to calculate by means of such a table as Table II. how much heat the amounts of these substances present in a pound of coal would produce if they were burnt separately, and then to assume that they produce exactly the same amount if burnt when united in the form of coal. To such a calculation a correction is ordinarily introduced. If the hydrogen or any part of it exists in the coal united with some of the oxygen, in the proportion in which these two substances combine to form water, it is, as it were, burnt before combustion takes place. It does not unite with more oxygen when the coal is ignited, and so does not help to increase the heat then produced. It is therefore usual to assume that of the hydrogen present in coal that only is useful in the production of heat which is in excess of the quantity

R

necessary to convert any oxygen the coal may contain into water.

An example will perhaps make this method clearer.

Let us suppose that a certain coal is found to contain

80·0 per cent. of Carbon
4·4   „   Hydrogen
3·2   „   Oxygen
12·4   „   Ash, nitrogen, &c.

Since oxygen unites with one-eighth of its own weight of hydrogen to form water, the 3·2 parts of oxygen would render ·4 parts of hydrogen useless for heat-producing purposes, and we should only have 4 per cent of *available* hydrogen. Hence (using Prof. Andrews' results) the heat produced would be :

By the Carbon :

$$14220 \times \frac{80}{100} = 11376 \text{ units}$$

By the Hydrogen :

$$60986 \times \frac{4}{100} = \underline{2439\cdot44 \quad ,,}$$

$$\text{Total } 13815\cdot44 \quad ,,$$

The number of units of heat produced by the combustion of a pound of coal is often called the calorific power of the coal. When calculated according to the above rules it is called the theoretical calorific power. Below are the theoretical calorific powers of two of the coals the analyses of which are given on p. 176. In each case the quantity of nitrogen present has been assumed to be 1·8 per cent., and as the analyses are of ash-free coal, the additional assumption has been made that the coals contained 5 per cent. of ash, and the numbers calculated from the Table on p. 176 have therefore been multiplied by ·95.

TABLE IV.

	Calorific Power
Bituminous Coal	13691·9
Anthracite	14285·9

This method of calculation is however open to considerable objection, the nature of which we shall perhaps best comprehend by considering in the first place some experiments made by Favre and Silbermann on the combustion of carbon.

This element is one of those substances which exist in several so-called 'allotropic' states. In all of these its chemical properties are precisely the same, though its appearance and physical properties are widely different. The diamond, black lead, and charcoal are all but various forms of pure carbon. No chemical tests can distinguish between them. The differences are due to the fact that though the atoms of which they are composed are the same in each, they are in each arranged among themselves in a different manner. Favre and Silbermann however found that the heats of combustion of these forms of carbon differed considerably. The following Table embodies their results.

TABLE V.

Substance	No of units of heat produced by its combustion
Wood Charcoal	14,544
Gas Retort Carbon	14,485
Native Graphite	14,035
Artificial Graphite	13,972
Diamond	13,986

These figures point to the conclusion that the heat of combustion of an elementary substance depends not only upon its chemical constitution but also upon its physical state before combustion. It varies both with the nature of the atoms and with the manner in which they are grouped together. We cannot deduce the calorific power of graphite from that of charcoal, nor that of the diamond from either. If then the mere fact that a sub-

stance is composed of pure carbon is not sufficient to
determine its heat of combustion, it is not reasonable to
suppose that the like information can be acquired in the
case of so complex a substance as coal, by a calculation
based only on a knowledge of the quantities of carbon,
hydrogen, and oxygen which it contains.

Coal is not a mere mixture of these elements, in which
they exist side by side, but separate and distinct the one
from the other. It is rather a mixture of their compounds, and when the coal is burnt any of these compounds
which are different from the products of combustion must
be broken up before the elements of which they are
formed can unite with the oxygen of the air. The heat
of combustion of the coal will depend in part upon the
nature of these preliminary decompositions. As a general
rule cold is produced when a compound body is resolved
into its elements, and if this holds in the case of those
substances which occur in coal, the true heat of combustion will be less than that calculated by the above
method. Part of the heat due to the combustion of the
carbon and hydrogen will disappear in making good the
loss which occurs before those elements are set free from
the bonds by which they have been hitherto united.
The rule however is not without exceptions, and if the
destruction of the compounds in question is attended by
the evolution of heat, the calorific power of the coal will
be increased instead of diminished by their presence.

In order therefore to calculate precisely the calorific
power of coal, it would be necessary to be precisely informed as to the details of its chemical composition, and as
to the quantities of heat which appear, or disappear, during
the complicated chemical transformations which take
place in the process of combustion.

Nor is this all; Prof. Andrews in common with all

other investigators determined the calorific power of hydrogen when that element was in the gaseous state. Till quite recently this gas, together with oxygen, nitrogen, and atmospheric air, had resisted all attempts to convert it into a liquid. The difficulties which had previously proved insurmountable, were however overcome almost simultaneously by two gentlemen, working independently of each other, and by very different methods. M. Cailletet first liquefied hydrogen in Paris on Dec. 31, 1877, M. Raoul Pictet not only liquefied, but also solidified it in Geneva on Jan. 10, 1878. Some idea of the magnitude of this achievement may be obtained from the fact that to attain the result the gas was subjected to a pressure of more than four tons to the square inch, and was cooled to 252° F. below the melting point of ice. To means such as these all the above-mentioned gases have now succumbed, and there remains but one (phosphorus pentafluoride) which has never been liquefied, and that will in all probability offer no special resistance whenever the attempt is made to condense it.

Investigations on the properties of liquid oxygen, hydrogen, nitrogen, and air will no doubt follow on these magnificent researches, but at present we are utterly ignorant of the properties of solid hydrogen, while of the liquid we know only that it is opaque and of a steel-blue colour.

We do however know that if a solid or liquid is converted into a gas its temperature will fall unless it is at the same time supplied with heat. Therefore even if the assumption involved in the above calculations, that the 'available' hydrogen is not combined with any of the other elements present in the coal, were correct, the calculations themselves would be open to objection. Such hydrogen would certainly exist in the coal in a solid state, and during the process of combustion would be converted into a gas. During this operation its temperature would tend to fall, and the heat spent in preventing this would be subtracted from the total heat of combustion of the coal. Prof. Andrews' experiments are not therefore really applicable to the case under consideration, and in assuming that the calorific power of solid hydrogen is, like that of gaseous hydrogen 60,986 units, we commit an error of the existence of which we are certain, while we are totally ignorant of its magnitude.

Nor are experimental proofs wanting to confirm the doubts which theory suggests as to the accuracy of this method of calculation. Two physicists, Scheurer-Kestner and C. Meunier, have of late years made a long series of experiments on the heat of combustion of coal. They analysed numerous specimens, calculated their calorific power by the ordinary rules, and then made direct experiments to determine their heat of combustion. A comparison of the numbers obtained by calculation and observation proved that they did not agree.

Thus in the case of two coals, the one from Ronchamp and the other from Creusot, which contained almost precisely the same proportions of carbon, hydrogen, and oxygen, the calorific powers instead of being, in accordance with the rules, identical, were 16,411 and 17,320 respectively. The difference between the real and calculated calorific powers amounted in some instances to as much as fifteen per cent.

In the case of two specimens of coal from England, and several from France, the calculated heat of combustion was too small. In that of six kinds of brown coal from France and Germany, it was too large, while experiments made on several different coals from Russia proved that in these cases the discrepancies between calculation and experiment were comparatively unimportant.

Another observer, L. Grüner [1] has also arrived at the result that the calorific power of coal cannot be calculated from a bare knowledge of its constituents. He inclines to the opinion that the heat of combustion becomes greater as the quantity of coke produced by the coal increases, but some of Scheurer-Kestner's experiments seem to negative this supposition.

Von Hauer [2] has investigated the relation between the calorific power and the geological age of Austrian coals. We should expect from the general nature of the chemical changes which take place in the conversion of vegetable matter into coal, that its calorific power would increase with age. The oxygen and nitrogen, useless as heat producers, gradually disappear, leaving behind a larger proportion of carbon. Von Hauer finds that this expectation is on the whole justified, though the difference between the calorific powers of two coals belonging to different geological epochs is often less than when the specimens compared are of approximately the same age.

On the whole then there appears to be no method of accurately determining the calorific power of coal, except by direct experiments on each particular kind. The investigator must assure himself by analysis of the ash and of the gaseous products of combustion that the coal has been completely burnt, and should this not be the case, must correct his result for the amount of combustible matter which remains free from, or is only partially combined with, oxygen. Such experiments require however a trained and skilful hand, and several methods have been suggested, by means of which it was hoped that the calorific power of coal might be determined with sufficient accuracy by experiments of a less refined and delicate

[1] *Jahresbericht Chem.*, 1187. 1874.
[2] *Jahrbuch für Min. Geol.*, 727. 1863.

nature. One of these, proposed by Berthier, consists in mixing the coal in a state of very fine division with about forty times its weight of litharge, or oxide of lead, and raising the whole to a red heat in a crucible. The lead gives up the oxygen with which it is united to the carbon and hydrogen of the coal. Carbonic acid and steam are given off, and a quantity of metallic lead is left behind, from the weight of which the calorific power of the coal is deduced. Experiment has shown that one part of pure carbon will reduce 34·5 parts of lead to the metallic state, and the calorific power of the coal is calculated by assuming that the amount of heat produced by its combustion is to that evolved when an equal weight of carbon is burnt, in the same proportion as the quantities of lead which they reduce. This calculation might perhaps serve in the case of a substance in which carbon was the only combustible element. Omitting for the moment such considerations as that some of the other constituents though not burnt might be vaporised, the heat of combustion would be proportional to the carbon burnt, and therefore to the lead reduced. If therefore the quantity of hydrogen in the coal is small, or if the hydrogen and oxygen are present in the proportions in which they form water, so that the hydrogen neither reduces the oxide nor produces heat when the coal is burnt, fairly good results are obtained. If these conditions are not fulfilled, a part of the lead —how much the experiment does not enable us to determine—is reduced by the hydrogen, and by assigning the whole to the carbon very serious errors may be introduced.

An instrument called Thompson's Calorimeter, which is represented in Fig. 51, is also sometimes used for determining the calorific power of coal. The coal is powdered and mixed with from ten to twelve times its weight of a

mixture of three parts of potassium chlorate and one of
nitre. This powder, which will burn without access to
the air, is placed in a copper cylinder, which in turn is
fixed in a larger vessel furnished with a stop-cock above

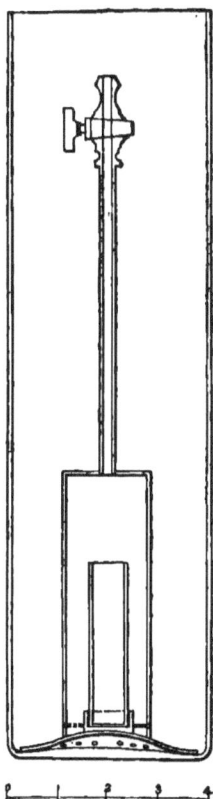

FIG. 51.—Thompson's calorimeter.

and pierced with holes below. The whole is immersed in
a known weight of water, and when the mixture is burnt
the hot gases bubble up through the holes and warm the
water. When combustion is finished, the stop-cock is
opened, the water fills the vessel, and the heat of combustion of the coal is deduced from the elevation in its
temperature. The quantities of coal and water to be used
with the instrument are so adjusted as to make the neces-

sary calculation extremely simple. Dr. Percy however
has proved that the bubbles of hot gas are not completely
cooled while rising to the surface, and that the loss of
heat thus caused is ' not unimportant.'

It is evident from all these considerations that many
of the experiments which have been made on the calorific
power of coal must be rejected. The method employed
by Scheurer-Kestner appears to be theoretically the best,
and a number of his results are given in the following Table.

TABLE VI.

Locality	Nature of Coal	Calorific power of dry Coal free from Ash
Toula, Russia	Lignite	13,837
Manosque, Basses Alpes . .	,,	12,584
,, ,,	,,	13,253
France and Germany . .	Brown Coal	11,340–14,220
England	Caking Coal	15,804
,,	,,	16,108
Basin of Donetz, Russia . .	,,	15,651
Creusot, France . . .	,,	17,319
,,	Anthracite	17,021
Basin of Donetz, Russia . .	,,	14,866

All these calorific powers are calculated for a pound
of matter really burnt, i.e. the ash, and any water which
may exist in the pores of the coal are supposed to be removed. After making allowance for these we may assume
that a pound of good coal, in the state in which it is
ordinarily used, gives out during combustion about 14,000
units of heat.

Another important enquiry must now be discussed,
viz. : What is the highest temperature that can be produced by the combustion of coal ?

This, like the last, is a practical question. It is a fundamental property of heat that it will only pass of its
own accord from bodies of higher to those of lower temperature. If therefore we wish to heat any substance,

e.g. the water in the boiler of an engine, or an ore which is to be fused, we must bring it in contact with something hotter than itself. If the gases of the furnace were cooled down to the temperature of the substance to be heated, they would cease to supply it with heat, and it is only the heat which raises them above that temperature as distinguished from that which brings them up to it that can be usefully employed.

It is therefore often important that a fuel should not merely supply a large quantity of heat, but that it should supply it at a very high temperature.

The greatest increase of temperature which it is possible to suppose the combustion of any substance could produce, would be attained if the heat of combustion were spent only in warming the products of combustion. If the nature of these products, the amount of heat required to raise the temperature of a pound of each through any number of degrees, and the calorific power of the substance burnt are known, this maximum increase of temperature can be calculated, and when so determined is called the theoretical calorific intensity of the fuel. Thus the theoretical calorific intensity of hydrogen is 15,802° F.; or in other words, if hydrogen were burnt in presence of the exact amount of oxygen necessary to convert it into water, both gases being at 32° F., the temperature of the steam produced could not possibly be greater than 32° + 15,802° F. If the hydrogen were burnt in air the theoretical calorific intensity would be only 6,570° F., as part of the heat produced would be spent in raising the temperature of the nitrogen which would then be mingled with the other gases.

Numbers such as these are however worthless for many reasons, of which it may be sufficient to refer to one. When a compound body is raised to a very high temperature a phenomenon known as 'dissociation' often occurs, i.e. a portion of the body is broken up into its elements. Thus if steam were strongly heated in a closed vessel, a part would be decomposed, and the vessel would be filled with a mixture of steam, oxygen, and hydrogen. The relative quantities of each of these substances would depend upon the temperature, an elevation in which would entail further decomposition of the steam, while a fall would be accompanied by a recombination of some of the oxygen and hydrogen. The three gases could not remain permanently in contact at any temperature above that at which dissociation first occurs, except when mingled in the proportions proper to that temperature, and to the pressure to which they were subjected.

Now if hydrogen is burnt in presence of the exact quantity of oxygen necessary to convert it into water, the heat produced by the combustion of a part of the hydrogen only is sufficient to raise the temperature of the mixture to such an extent that no further union of the elements can take place. As soon as the temperature begins to fall more water vapour is formed, and each step in the process of cooling is thus accompanied, and of course retarded, by the combustion of fresh quantities of hydrogen until the whole is consumed. The rapidity with which all these phenomena succeed one another is almost inconceivable. Thus from some experiments which Bunsen performed, he calculated that a mixture such as that above referred to, when exploded by an electric spark, reached its maximum temperature in the vessel which he employed in $\frac{1}{4000}$ of a second, while the burning gas only emitted a bright light for $\frac{1}{65}$ of a second. In that short time, then, a large quantity of the heat produced had been dissipated; but had it been possible to prevent this loss, the very intensity of the heat would have checked the process of combustion, and would thus have made it impossible for the temperature of the mixture to have risen to that given by the theoretical calorific intensity of hydrogen.

Bunsen attempted to determine by experiment the maximum temperature actually attained when hydrogen and carbonic oxide were burnt in presence of greater or less quantities of oxygen, or of air. His calculations involved some doubtful assumptions, but a French chemist, Berthelot, has recently shown that Bunsen's data can be used to fix two limits between which the maximum temperature must lie. The following table gives (1), the theoretical calorific intensity of hydrogen and carbonic oxide; (2), the maximum temperature produced by the combustion of those gases according to Bunsen; (3), the limits to that temperature fixed by Berthelot. The quantity of oxygen present is supposed to be the amount necessary to convert the hydrogen into water or the carbonic oxide into carbonic acid whether the gases are burnt in oxygen or in air, and the volume of the gases to remain constant during combustion.

TABLE VII.

	Calorific intensity in degrees Fahrenheit		
	(1) Theoretical	(2) According to Bunsen	(3) Limits according to Berthelot
Hydrogen burnt :			
(1) in Oxygen . . .	15,802°	4,954° [1]	6,773°–4,354°
(2) in Air	6,570°	3,503° [1]	3,827°–3,087°
Carbonic Oxide burnt :			
(1) in Oxygen . . .	16,104°	5,458°	7,236°–4,633°
(2) in Air	7,268°	3,593°	3,857°–3,210°

[1] These numbers are a little below those given by Bunsen, a correction (for the latent heat of steam), which he omitted, having been introduced.

The figures in this table prove that we are unable to

calculate with accuracy the maximum increase of temperature attainable when hydrogen and carbonic oxide are
burnt. When however they are consumed in air, the
possible limits to that increase are not wide apart, and
as the extremes fixed by Berthelot are almost certainly not
very near the truth, Bunsen's results of 3,500° and 3,600°
F. may be provisionally accepted as correct.

If coal is completely burnt the products of combustion are for the most part the same as those formed during
the combustion of these gases, but it does not follow that
the real calorific intensity is the same in both cases, and
several causes would tend to make that of coal the smaller
of the two. The quantity of heat produced when carbon
is burnt in air is greater than that generated when a
quantity of carbonic oxide containing an equal amount
of carbon is consumed, but as double the amount of
oxygen, and therefore also of nitrogen, has to be supplied
in the former case, the total amount of matter to be
heated is greater, and the theoretical calorific intensity is
6,801° F., or somewhat below that of carbonic oxide. All
the above numbers in Table VII refer also to gases which
are kept while burning at constant volume, whereas
when coal is burnt in a furnace the gases are free to
expand, and by expansion, as we shall see later on, they
become cooled, and thus require from 30 to 40 per cent.
more heat to raise their temperature through a given
number of degrees.

These causes may be partly neutralised by a more
complete combustion of the carbon. According to Bunsen
when hydrogen and carbonic oxide were burnt in oxygen,
the maximum temperature was reached, when about one-
third of the total quantity of combustible gas present was
consumed, but this fraction was increased to about one
half when the same gases were burnt in air, and it may

be further augmented when the proportion of nitrogen to carbon is still larger. There is also some direct evidence that the temperature of burning gases when they are allowed to expand is not so much below that attained when the volume is kept constant as we might at first expect. Thus Deville found that the temperature of the oxyhydrogen flame is 4,500° F., a result which from the circumstances of the experiment is probably too low. In his experiment the gases were free to expand, while according to Bunsen the temperature reached when the volume is constant is only 4,950° F. On the whole then though we are unable to fix with any great precision on the maximum temperature attainable by the combustion of coal, we are assured that the so-called theoretical calorific intensities of fuels are numbers of no practical or scientific value and calculated only to mislead. The elevation of temperature actually produced in practice differs very much in different fires. A writer on the blast furnace (Schinz) describes these variations as enormous. He estimates the average temperature in the hearth of a blast furnace at about 3,800° F., but in some instances his calculations point to a maximum temperature of from 800° to 900° higher. This, however, if really attainable, is only reached by employing fuel which has itself been raised to a very high temperature.

In conclusion, it must be remembered that the fitness of coal as a fuel for any particular purpose often depends not only on the amount and temperature of the heat which it can supply, but also on its physical properties, its chemical composition, and its behaviour while burning. Thus since space is very valuable in steamships, it is necessary that in them the coal should be packed as closely as possible. The space occupied by a given weight of coal will depend in part upon its density, and in

part upon the shape of the fragments into which it is
broken, and the latter will in turn depend upon the natural
cleavage of the coal or the mutual inclination of the
directions in which it breaks most easily. Different coals
vary much in this respect, some occupying 20 per cent. more
room than others. The preference is given to those
which break up into fragments most nearly approximating
to the form of a cube.

The presence of sulphur in coal renders it unfit for
certain metallurgical processes, and the character of the
flame and the quantity and nature of the ash produced
often determine whether it is or is not adapted to the
special requirements of the case.

Again the amount of heat which is extracted from any
particular coal depends as much upon the skill of the
stoker, and the construction of the furnace, as upon the
absolute calorific power of the fuel. More heat may be
lost than gained by the employment of a coal with
a high heat of combustion in a furnace which it does not
suit. The choice between different coals can only be made
with absolute certainty after they have been tried in the
exact circumstances under which they are to be employed.
Many attempts have been made to estimate the value of
coal as a fuel by determining the weight of water evaporated
by a given weight of the coal when burnt in the furnace
of a steam engine. The number of pounds of water at
212° which a pound of coal can convert into steam at the
same temperature is called its evaporative power. If the
heat of combustion is 14,000 the evaporative power is 14·5.
The results of such experiments cannot however be considered as applying to furnaces and boilers dissimilar to
those actually used, and in some cases they may lead to
very erroneous conclusions.

It is evident from the considerations which have been

adduced in this chapter that science in the workshop as in the laboratory must be founded on frequent and often repeated experiment. It would be a mistake to suppose that the cultivation of science is worthless because we cannot always in the present state of our knowledge predict exactly what will be the results of a given combination of circumstances. That indeed is the goal of our endeavours, but at present we must often be contented to be wiser than our predecessors only in our knowledge that certain causes are at work, though we are unable to calculate their effects.

We are not able to deduce the calorific power and intensity of coal from a knowledge of its chemical composition, but we at least know the grounds of our inability, and something of the nature of the problems the solution of which would remove it. Nor is there any ground for the belief that the difficulties which at present bar our way are insurmountable. The science of thermo-chemistry in which chemical change is studied with reference to the alterations of temperature by which it is generally attended is of recent origin; but in this direction, as in others, physics and chemistry are advancing to meet one another, and we may expect from their union results which could never have been attained by either alone.

258

CHAPTER VIII.

COAL AS A SOURCE OF POWER.

IN the last chapter we discussed the principles on which
the measurement of heat is founded, and applied them to
the determination of the quantity of heat produced by the
combustion of coal. We must now briefly consider the
theory and practice of the employment of this heat as a
source of mechanical work.

Work may be measured by comparing it with the labour
which must be expended in order to raise a pound through
the height of a foot. The amount of work which is performed in accomplishing this task varies with the place at
which it is undertaken, as the force with which the earth
attracts a pound of matter is different at different places.
These variations however are but small, and may for our
present purpose be neglected. We may consider the elevation of a pound through a foot as involving a definite
amount of work, which we select as our unit, and to which
we give the name of a 'foot-pound.' If a weight other
than a pound is raised through a distance greater or less
than a foot, the work spent in the operation is calculated
by multiplying the weight expressed in pounds by the
distance expressed in feet; thus to raise $5\frac{1}{2}$ pounds through
18 inches ($1\frac{1}{2}$ feet) requires the expenditure of $5\frac{1}{2} \times 1\frac{1}{2}$ or
$8\frac{1}{4}$ foot-pounds. Similarly if a locomotive pulls a train
for a mile, always exerting upon it a force just equal to

the weight of a ton and a half, the work which it performs
is expressed in foot-pounds by multiplying the number of
pounds in a ton and a half by the number of feet in a
mile. The foot-pound can thus be employed to measure
work of all kinds, and having decided to employ it as our
unit, we may now pass to a consideration of the circumstances under which bodies possess the power of performing work.

To this power a special name is given, *energy*, and the
first point that demands our attention is the fact that
moving bodies possess energy in virtue of their motion.

Let two unequal weights be connected by a string
which passes over a smooth pulley. If the length of the
string is such that when the larger weight rests upon the
ground the smaller hangs suspended in the air, it is evident
that the smaller weight, when at rest, will be utterly incapable of raising the larger through the smallest fraction
of an inch. If however it be lifted and then allowed to
drop, it will soon be arrested by the string, but before it
is brought to rest it will with a jerk raise the larger
weight from the ground. The distance through which
this is raised will depend in part upon the relative mag-
nitudes of the weights, and in part upon the distance
through which the lesser one has fallen, but however
small it may be, the fact that the larger weight is moved
at all proves that a body possesses when in motion a
power of overcoming resistance which it does not possess
when at rest. Further illustrations of this truth, if
needed, are not far to seek. A nail is driven into a block
of wood, not by piling weights upon it, but by striking it
with a swiftly moving hammer. The greatest possible
velocity is imparted to projectiles intended to pierce the
armour of a hostile ironclad or fort. In railway accidents
it is not only the great weight, but also the high speed of

the trains which enables them to shatter themselves, and any object with which they may come in contact.

Taking it then for granted that moving bodies possess the power of doing work, let us enquire what becomes of this energy when their motion is arrested. Let two ivory balls be suspended from a nail by threads of equal length, and let matters be so arranged that they can swing backwards and forwards in front of a graduated scale. If one ball be pulled aside through several divisions and then allowed to swing, it will strike the other, which will move on, while the first will be brought to rest (or very nearly so) at the point where the two meet. The scale will make it easy to prove that the second ball will move away from the central point through as nearly as possible the same number of divisions as the first had moved through towards it.

In this experiment we observe that the energy of the moving ball is not destroyed when it is stopped. Had it been allowed to swing on, it would have been able to lift itself to the same height as that from which it had fallen, and when itself arrested it transmits to its fellow the power of performing precisely the same feat. It imparts to another body the energy of which it is itself deprived. This is an example of the truth of the doctrine known as the *conservation of energy*, which asserts that energy is as indestructible as the matter in which it resides, and that, though it may change its seat and its form, it can never be either created or annihilated by any force or power with the action of which we are acquainted.

A single experiment is however but a slender basis on which to rest so wide a generalisation, and objections to the doctrine may at once be raised. A ball thrown vertically upwards after a while comes to rest in the air, when, having ceased to rise, it is on the point of falling. Under these circumstances its own motion has been lost, and yet it has communicated motion to no other body. Has not energy in this case been destroyed? To this enquiry we must answer—no. The ball at the highest point of its course has—what it had not when on the ground—the power of falling. Its energy of motion has it is true disappeared, but in losing it, it has attained a position from which by retracing its path the lost motion can be regained. For the infinitely short space of time during which it remains at rest it possesses the power of doing work in that it possesses the power of acquiring motion.

Consider next the case of a falling body which strikes the ground and is suddenly brought to rest; or a pendulum which at the lowest point of its course meets, not another, but an obstacle, such as a wall, which destroys its motion without itself being moved; or a train which is stopped by the action of a break. In all these cases the moving body is rendered motionless, yet it has gained no position of advantage for reacquiring motion, nor has it set any other body in movement. What then has become of the energy which it once possessed? The answer to this question is by no means obvious, but it is evident that any attempt to reply to it must be preceded by a careful examination of bodies, the motion of which has been destroyed under these, or similar circumstances. By this means only can we determine whether any change, either in themselves or in their surroundings, will account for the missing energy.

If instead of employing a light ivory ball, as in the experiment just described, a heavy weight, say 14 lbs., be suspended by a cord, and be allowed to swing against a still heavier weight placed upon the ground, it will be

found that if a small piece of phosphorus be placed at the
point where the two meet it will be ignited by the blow.
Phosphorus, though very combustible, yet like most other
substances, requires heat to set it on fire; and this experiment proves that the destruction of the motion of a
pendulum, when that motion is not transmitted to another
body, is accompanied by the appearance of heat.

Let us take another example. Fig. 52 represents a
small piece of apparatus consisting essentially of a wooden

FIG. 52.—Apparatus for illustrating the production of heat by friction.

wheel, which can be set in rapid rotation, and a break by
which its motion can be checked. The break, which is
applied by a lever, is itself a hollow brass box, through
the top of which passes a curved glass tube. The box
and part of the tube are filled with coloured ether, and
the whole thus constitutes a delicate thermometer, a magnified image of which, together with that of part of the
wheel, can, if the experiment is to be shown to a large
audience, be thrown upon a screen. The rapid expansion
of the ether as soon as the break is applied, proves that

when the rotation of the wheel is retarded, the break and its contents are warmed.

These experiments teach us that, in the cases we have studied, the destruction of visible motion is accompanied by the appearance of heat. The two phenomena occur simultaneously, and it seems, even at first sight, probable that they are related as cause and effect. We must regard the heat which is created as in some manner the equivalent of the energy of the visible motion which is lost. Nor is there any difficulty in this, if we once admit that heat itself is energy of motion of a peculiar kind. The true explanation of what takes place when bodies are arrested under circumstances such as those above described is, that they do not in reality stop moving but only commence themselves to move, or to set other bodies moving, in a different way. The heat which they manifest is due to a vibratory motion of their individual particles, and this is increased in the case of the pendulum by the jar of the blow by which it is brought to rest, in that of the wheel and break by the friction between them.

A description of a simple experiment may perhaps make this clearer. Let a couple of metal hoops to which a number of small bells are attached be suspended by a string. They can then be swung backwards and forwards, and the bells may be considered as rough and enormously magnified representatives of the individual atoms of which all matter is composed. If the hoops are allowed to swing undisturbed, no sound is emitted by the bells. The pendulum as a whole moves, but its several parts vibrate either not at all, or but very slightly. It represents a cold body. Let now its motion be suddenly arrested in the middle of a swing. Instantly all the bells clang. The destruction of the motion of the body as a whole gives rise to a new vibratory movement of its parts. We are

informed of the existence of this movement by the tinkle
of the bells; the evidence of the analogous motion of the
real atoms of a body is the production not of sound, but of
a rise in temperature.

It is now possible to advance a step further.

We have seen that moving bodies possess the power
of performing work, and this, which is true of masses of
matter, will also be true of their individual parts. The
visible motion of a body as a whole, and the invisible
motions of its particles, alike constitute stores of energy,
differing only in the character of the movements, and in
the effects which they produce upon our senses. We can
by appropriate mechanism make either a moving or a hot
body do work, but in so doing we must deprive the
former of a part of its motion, and the latter of a portion
of its heat.

It is not difficult to prove that bodies may thus expend
their heat in the performance of work. Let some air be
forced into a metal vessel furnished with a stop-cock, and
let the whole be left for some time to acquire the temperature of the room. If the stop-cock be then opened
the compressed air will rush out. When the efflux has
completely ceased, let the cock be closed, and the vessel
be left undisturbed for some minutes. On once more
opening the cock a fresh rush of air will take place from
the reservoir. This is explained by the fact that the air
left inside the vessel when the stop-cock was first opened
and closed had in expanding helped to push out the rest
of the air. To do this work it drew upon its own stores
of heat and thus became cold. In this state it was once
more imprisoned, and though it recovered its lost heat
from the bodies in its immediate neighbourhood, it was
unable to expand. When the cock was again opened
the expansion proved that the air had grown warmer

since it was enclosed, and therefore that at the moment it was shut in it must have been colder than surrounding substances, i.e. colder than itself had been before it had been made to perform work.

If then we accept the doctrine that heat is a form of energy, that it can be produced by work, and that on the other hand we can employ heat to set bodies in visible motion, to raise weights, or perform any other mechanical task, it becomes a matter of great interest to determine

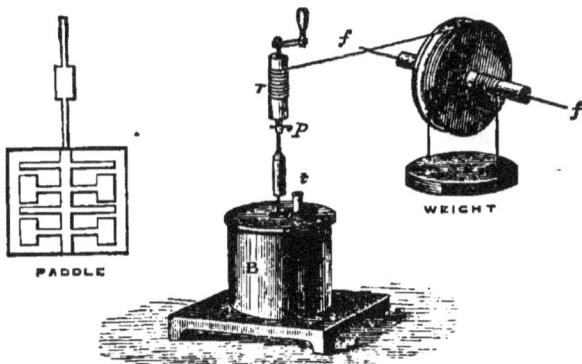

FIG. 53.—Toule's apparatus for measuring the mechanical equivalent of heat.

how much heat or work can be obtained by the expenditure of a given amount of work or heat. The first person who attacked this question with a clear conception of the experimental and theoretical problems which its solution involved, was Dr. Joule of Manchester. His earliest paper on the subject was read before the British Association in 1843, and the result of his experiments was finally communicated to the Royal Society in 1849.

The apparatus which he used is represented in Fig. 53. It was designed to enable him to perform a known amount of work, to spend this work in producing heat, and to measure the amount of heat generated.

The work was performed by two weights (only one of which is shown in the figure) which were allowed to fall through a measured height. They were suspended by strings which were first rendered horizontal by passing over pulleys so mounted as to move with very·little friction, and then wound round a vertical wooden cylinder (r). This cylinder could be easily attached to or detached from a vertical rod, the lower end of which was furnished with eight projecting vanes or paddles and was immersed in a vessel of water (B).

When an experiment was performed the cylinder (r) was detached from the rod and held by a wooden support, so that the weights were wound up without disturbing the rod, the paddles, or the water. The cylinder was then attached to the rod, and the weights allowed to fall. When the weights were lifted a certain amount of work was spent in conferring upon them the power of falling. Had they been allowed to fall freely, they would, the instant before striking the ground, have exchanged this power for an amount of energy of visible motion just sufficient to enable them to perform the same amount of work as had been spent in raising them. On striking the ground this would have been converted into the energy of heat. The weights, however, were not allowed to fall freely; their motion caused the paddles to revolve, and the rotation of these was checked by the water in which they were placed. The weights thus arrived at the ground with but a small velocity, the energy they would otherwise have possessed had been spent in imparting motion to the paddles and the water, which motion, having been destroyed by the friction of the water against itself and against the sides of the vessel, had been converted into heat. To assist in this transformation the vessel was furnished with barriers which arrested the motion of the liquid, and which were so shaped that the paddles could only just pass them. They are shown in section in Fig. 53.

The experiment of raising the weights and allowing them to fall was repeated twenty times, and the rise in the temperature of the water was noted. The necessary allowances were then made for the facts that the vessel and paddles were warmed together with the water, that the weights reached the ground with a certain small velocity, &c., and all such corrections having been introduced, the experiment furnished the means of calculating the amount of heat produced by the expenditure of a given amount of work. The result at which Dr. Joule arrived was that each unit of heat was produced by the expenditure of 772·69 units of work.

Many other observers have since attacked the same problem by various experimental methods. Some of these were from the first unlikely to give as good results as those which Dr. Joule had obtained, but it was very important that the determination of the number of foot-pounds necessary to produce a unit of heat should be undertaken in as many different ways as possible.

The mean of sixteen of the most accurate of these determinations gives 786 as the value of the mechanical equivalent of heat.

An extremely careful series of experiments by Prof. H. F. Weber, of the Federal Polytechnic Academy at Zurich, has lately been published. The value at which he arrives is 780 foot-pounds. This number, which is obtained by a method essentially different from that just described, and depending upon electrical measurements, is supported by that (782) deduced from the properties of air.

Dr. Joule also has himself just redetermined the value of the mechanical equivalent of heat by a method analogous to, but in some respects differing from, that which he formerly employed. The result of his latest experiments was only communicated to the Royal Society

in November last (1877), and is, in his own words, 'that
taking the unit of heat as that which can raise a pound of
water weighed *in vacuo* from 60° to 61° of the mercurial
thermometer, its mechanical equivalent reduced to the
sea-level at Greenwich is 772·55 foot-pounds.'

The wonderfully close agreement between this
number and that obtained by Dr. Joule nearly thirty
years ago must, if it were possible, lead the scientific
world to place a still higher estimate on his work than
before. It is impossible at present to compare his results
exactly with those of Weber and others, as different
observers have unfortunately used slightly different units
of heat. Before long it is hoped that data may be forth-
coming by the aid of which such comparison may be made.
Dr. Joule's result of 1849, confirmed by that of 1877,
will however always rank as a marvel of experimental
skill. During thirty years of active criticism and rapid
advance in all departments of knowledge, it has maintained
its position as by far the best determination of perhaps
the most important constant in the whole range of science,
and Dr. Joule's resolve himself to test his earlier work has
led to a result agreeing to within one five-thousandth part
of the whole with that which he first gave to the world a
generation ago.

These brilliant researches then inform us of the fact
that (in round numbers) 772 foot-pounds of work are necessary to produce a unit of heat, and, conversely, that a unit
of heat, if entirely used up in the production of mechanical
work will perform 772 foot-pounds. Let us now combine this result with that at which we arrived in the last
chapter as to the calorific power of coal. If we multiply
the number of units of heat produced by the combustion
of a pound of coal by 772, we obtain the number of
foot-pounds of work to which that heat is equivalent.

The following table gives the mechanical value, calculated by this means, of the coals referred to in Table IV.:

TABLE VIII.

Kind of Coal	Mechanical value in Foot-pounds
Bituminous Coal	10,570,147
Anthracite	11,028,715

If we take as before the average calorific power of a pound of good coal as 14,000 units, the average mechanical value of a pound of coal is 14,000 × 772 or 10,808,000 foot-pounds—in round numbers 10,000,000 foot-pounds. The following table may assist us to obtain a clearer idea of what these figures mean. It gives the number of foot lbs. of work which can be done under favourable conditions by a man and a horse, together with the number of pounds of coal the combustion of which would produce the same amount of work. The result is that a man and a horse, under the most favourable circumstances referred to in the table, could only do in a day as much work as is locked up in two-tenths and twelve-tenths of a pound of coal respectively.

TABLE IX.

Kind of Work	Foot pounds of work per day	No. of lbs. of Coal the heat of combustion of which is equivalent to work
Agent Man		
Raising his own weight up stair or ladder .	2,088,000	0·21
Carrying weights up stairs and returning unloaded	399,600	0·04
Pushing or pulling horizontally . . .	1,526,400	0·15
Working pump	1,188,000	0·12
Agent Horse		
Drawing cart or boat walking . . .	12,441,600	1·24
Drawing light railway truck (cantering and trotting)	6,444,000	0·64

We have now reached the end of the second stage of
our enquiry; we have determined the value of coal as a
source of power. We are therefore prepared to enter
upon the discussion of the use which we are making of
the stores of energy which are in that mineral placed at
our disposal. Let us in the first place enquire whether
the laws at which we have arrived by the aid of carefully

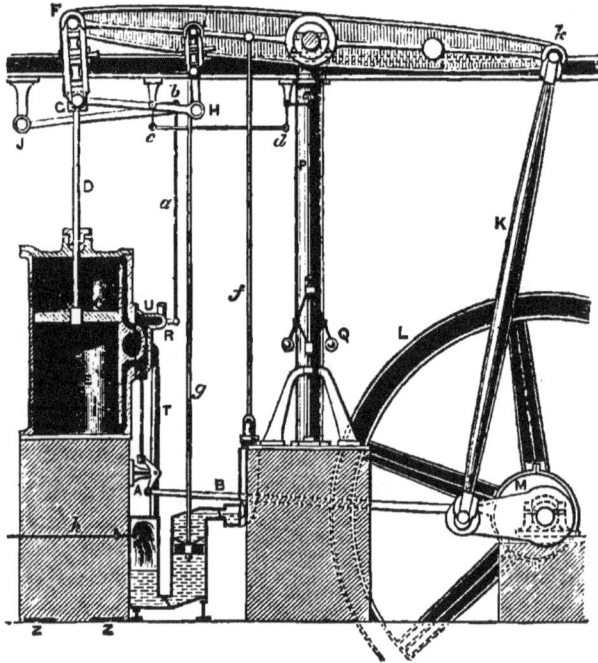

FIG. 54.—Steam Engine.

devised apparatus are really applicable to the rougher
machines employed for practical manufacturing purposes.

Is it, for instance, true that in the steam engine, heat
disappears when work is performed, and that each missing
unit of heat is accounted for by 772 foot-pounds of work?
This question has been attacked by M. Hirn, a French
engineer. In the steam engine heat is supplied by the
furnace to the steam and water in the boiler (Fig. 54), the

steam carries this heat with it into the cylinder where it pushes the piston o forwards, and thus imparts motion to the machine and enables it to accomplish work. At the proper time the communication between the boiler and the cylinder is interrupted, and at once or shortly afterwards a valve closing a pipe which connects the cylinder with a vessel, into which a stream of cold water is continually passing from the pipe h, and which is called the condenser, is opened. The steam is condensed, the piston moves back and ejects the remaining vapour and liquid into a vessel, to which they carry any heat they may contain. In a double acting engine such as that shown in the figure, the steam from the boiler is entering the space behind the piston, while that which entered during the last stroke is being driven before the piston into the condenser. The piston is thus always pushed by the steam in the direction in which it is moving. These details do not however for the moment specially demand our attention. The point before us is that the steam which is generated in the boiler ultimately reassumes the liquid form in the condenser, and will therefore, unless it has lost any heat by the way, carry with it to the condenser all the heat which was communicated to it in the boiler. The object of M. Hirn's experiments was to determine whether these two amounts of heat were equal, and if not, to determine the relation between the missing heat and the amount of work which the engine had in the meantime performed.

The experiments were surrounded by numerous difficulties, and I must content myself with indicating the principles of the method. The labours of M. Regnault have informed us of the quantity of heat necessary to transform a given weight of water into steam at a given temperature. M. Hirn knew the temperature of the steam in the boiler, and he also knew the quantity of water which was transformed into steam per minute. This was given by the quantity of water which it was necessary to supply to the boiler in each minute in order to keep the water-level stationary. A stream of cold water was kept flowing through the condenser, and this was so adjusted that when the machine was working regularly the temperature of the water in the condenser remained constant. Under these circumstances the heat given up by the vapour and liquid discharged into the condenser in a minute, was exactly the quantity necessary to heat up the cold water supplied in the same time to the temperature of the condenser. The amount of this water and its rise of temperature on entering the condenser being known, the quantity of heat given up by the steam was thus determined.

The amount of work performed by the engine was registered by an instrument invented by Watt, an improved form of which is shown in Fig. 55. A pipe in connection with the cylinder allows the steam contained in it to press upon a small piston. The pressure of the steam is thus measured by the distance through which it compresses a spiral spring which opposes the motion of the piston, and this is registered by a pencil attached to a lever, the movements of which are regulated by those of the piston. The point of the pencil is pressed against a sheet of paper, rolled round a cylinder which is made to rotate by a string attached to the machinery. The pressure of the steam is registered by the vertical movements of the pencil; the distances through which the piston moves under the influence of any particular pressure, are indicated by the movements of the cylinder, which tend to make the pencil trace out horizontal lines on the paper. The combined motions of the pencil and cylinder produce a curve from which the amount of work performed by the engine can be deduced.

The result of M. Hirn's experiments may be briefly summed up thus : (1), the steam returned less heat to the condenser than it had received in the boiler ; (2), when due allowance had been made for the heat given up in the various parts of the engine to the air and other surrounding substances colder than the engine itself, a con-

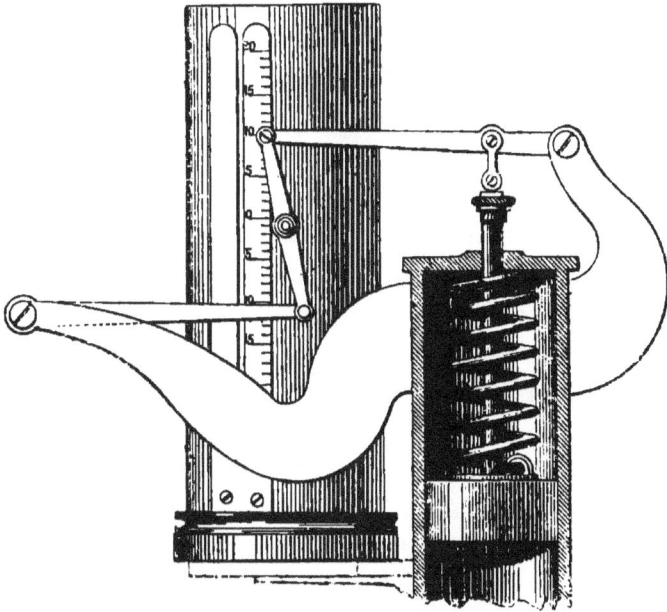

FIG. 55.—Indicator.

siderable quantity of heat was still missing ; (3), for every unit of heat so lost 716 foot-pounds of work had been done.[1]

The conclusion to be drawn from these experiments is obvious. Their difficulty rendered it from the first improbable that they would prove that the work performed and the heat lost were in the exact proportion indicated

[1] This number is, according to Verdet, more correct than that which is usually given as the result of these experiments, viz. 752 foot-pounds or 413 kg.-mètres. See *Œuvres de Verdet*, vii. 53.

by Dr. Joule. The most that could be expected was an approximation, and that was obtained. Our belief however that the laws we have discussed are applicable to the steam engine, is not based on the result of these direct experiments alone. Those laws have been tested in numerous instances in which more accurate measurement was possible, and they have always been found correct. It is doing no injustice to M. Hirn to say that the evidence afforded by his direct experiments is less convincing than the indirect evidence which is furnished by experiments in nearly every branch of science.

If then we admit that in the steam engine the heat developed in the furnace is the true source of the power of the machine, our previous calculations inform us of the fact that if the heat produced by the coal were entirely utilized, each pound of coal consumed would enable us to perform 10,000,000 foot-pounds of work. In reality but a small proportion of the total heat supplied by the combustion of the coal is employed in the production of useful work. The loss of the remainder is in part due to the imperfection of our machines. Every improvement is accompanied by a saving of coal, and further improvements in the same direction may no doubt be expected in the future. The theory of heat however unfortunately leaves no doubt that there is a limit to this process, and that a large part of the heat supplied to the engine must, whatever improvements we may effect, be lost for all practical purposes.

In the furnace, some heat is lost by the non-combustion or imperfect combustion of a part of the fuel. Some of the ash often consists of unburnt coal, and unless the supply of air is properly regulated, a certain amount of carbonic oxide instead of carbonic acid is formed. In order to prevent this it is found necessary to supply the furnace with a much larger amount of air than (generally about twice as much as) that which contains just enough oxygen for the combustion of the fuel.   This increased amount of air is in itself a source of loss.   The excess of oxygen and the large quantity of nitrogen swept through the furnace have to be heated, and thus keep down the temperature.   The heat therefore passes less readily from the furnace to the boiler than it would do if the difference of temperature were greater.   The strong current of air must also be kept up by some means.   If by a chimney, the gases as they leave the furnace must be hot to create the draught, and the heat they contain is lost.   If by a fan, the heat necessary to work it must also be considered as a loss.   A part of the heat produced may also be conveyed away by conduction through the brickwork, or by radiation from the furnace doors.

All these losses are theoretically preventable, and though they will probably never be completely done away with, those due to imperfect combustion, conduction, and radiation may, by careful firing and a properly constructed furnace, be rendered very small.   The net result however is that a considerable portion of heat is always lost in the furnace, and that from 50 to 70 per cent. only is transmitted to the water and steam, and that of this a small fraction only is converted into work.

As we have already seen, the steam, when it leaves the cylinder, having done its work, carries with it a considerable proportion of the heat which it received in the boiler. In a high pressure engine, such as a locomotive, where the spent steam is discharged into the air, this heat is irretrievably lost.   In an engine furnished with a condenser, a part may be usefully employed in heating the water with which the boiler is supplied, or by special arrangements in other ways.

Under all circumstances however a loss occurs; and
theory shows that however perfect a heat engine may be
some such loss is unavoidable. Into details of a proof of
this statement it would not be advisable to enter now.
Suffice it to say that it is based upon the fact that every
engine in which heat is continuously transformed into
mechanical work must work in a cycle, i.e. must perform a
series of operations at the conclusion of which the various
parts of the machine are in the same relative positions as
at first. This is the case in the steam engine which goes
through precisely the same set of movements during every
stroke, so that at the end the arrangement of the mechanism is the same as at the beginning. In order that a heat
engine may perform useful work and yet return after a
time to its original state, it is necessary:

(1). That it shall be supplied with heat at some period
of its stroke.

(2). That a smaller quantity of heat shall be abstracted
from it at another period.

(3). That the temperature of the heat supplied shall on
the whole be higher than that of the heat which is
abstracted.

This last condition makes it impossible to employ
again in the machine the whole of the heat removed from
it. We cannot make a cold body supply heat to a body
hotter than itself, and we cannot use the waste heat given
out by the cold parts of the machine to heat the warm
parts. If the spent steam from the cylinder of an engine
were discharged into the boiler, it would cool not heat the
contents. A loss is thus unavoidable, but it becomes less
as the range of temperature through which the machine
works increases. Thus in an engine in which the temperature of the steam when admitted to the cylinder is
250° F., and the temperature of the water in the condenser
is 100° F., it would be impossible to utilise more than 20

per cent. of the heat; but if the steam were heated at
400° F., it would be theoretically possible to employ 35 per
cent. Practical difficulties, which will suggest themselves
to every one, prevent the employment of either very high or
very low temperatures in the steam engine, and the imper-
fections of the machine reduce its efficiency considerably
below the amount theoretically possible, even for temperatures (such as those above cited) which are attainable in
practice.

The following table (based principally upon calculations
by Prof. Rankine), illustrates the large saving of heat
which has been effected in the past by improvements in
the steam engine.   In the first column is a description of
the engine ; in the second the number of units of heat
out of every 100 produced which are spent in generating
steam; in the third the number of units out of every
hundred which are converted into mechanical work.

The first four rows illustrate the change in the efficiency of a particular low pressure engine, supposing that
it was successively fitted with a steam jacket to the cylinder,
an apparatus for superheating the steam, and a Siemen's
Regenerator.

The fifth row gives the amount of heat utilised in the
engines of H.M.S. ' Briton,' which Mr. Bramwell has lately
stated to be the most efficient with which he is acquainted.[1]

TABLE X.

100 units of heat supplied.

	Heat trans-mitted to Boiler	Heat used in doing work
Unjacketed	54	5·9
Jacketed	54	6·6
Steam superheated	60	8·7
Working with Siemen's Regenerator . .	60	10·0
Engines of H.M.S. ' Briton '. . .	—	11·1

[1] Science Lectures at South Kensington, 1877. Macmillan and Co.,
The Steam Engine, p. 61.

These figures prove that in the best engines only about 11 per cent. of the heat produced is converted into work, and as a portion of the work done by the machine is spent only in overcoming the friction of its various parts, we shall not be far wrong in assuming that even in favourable cases more than 90 per cent. of the heat is wasted. The enormous loss of power which thus occurs in the steam engine naturally gives rise to the enquiry whether no other form of machine can be devised which shall be equally serviceable for practical purposes, but more efficient in utilising the heat of the fuel employed.

Without attempting to answer this question in full, let us briefly compare the steam engine in this respect with two or three other motors which are, or have been used for industrial purposes. In the air engine a piston is moved by the expansion and contraction of a mass of air which is alternately heated and cooled; the principal difference between it and the steam engine lying in the employment of a gas which is not liquefied at the lowest temperature to which the machine works, instead of a substance which is in turn liquefied and vaporised.

Table XI. contains the outcome of Rankine's calculations on two different kinds of air engines. The first, Ericsson's, was employed to propel a steam ship named after the inventor of the engine, the second, Stirling's, 'worked for several years at the Dundee foundry.'

TABLE XI.

100 units of heat supplied.

	Heat transmitted to air	Heat used in doing work
Ericsson's Air Engine	40	10·5
Stirling's ,, ...	44	13·3

On comparing these figures with those in Table X., it will be seen that in these engines a very large proportion of the heat actually passed into the air is utilised, but that, owing to the extreme difficulty with which air takes up heat, only a very small proportion of the heat produced reaches it. Still in spite of this drawback Stirling's machine was more efficient than the best steam engines. Air engines have not however been much employed, partly because they are generally of an inconvenient bulk, and partly because the hot air oxidises and thus destroys the metallic surfaces with which it is in contact.

Another form of motor now coming into general use is the gas engine. In this machine a mixture of air and coal-gas is exploded by an electric spark, or by contact with a flame. The explosion forces up the piston, and when the work is done the spent gases are discharged into the atmosphere. Calculations which have been made on the supposition that no heat is communicated by the gases to the piston and cylinder indicate that such machines should convert a very large percentage of the heat supplied into work, whereas experience proves that they are in this respect not superior to the steam engine. The discrepancy between theory and practice is no doubt in part due to the enormous loss of heat which must take place owing to the high temperatures attained by the products of combustion, but the two can be brought into somewhat closer agreement by allowing, however roughly, for the effects produced by dissociation. Thus M. Verdet ('Théorie Mécanique de la Chaleur,' ii. p. 239) finds that a theoretically perfect gas engine, in which the explosive mixture consisted of carbonic oxide and air containing just enough oxygen to convert the carbonic oxide into carbonic acid, would convert ·408 of the heat of combus-

tion into work ; a result which, if correct, would make it
one of the most perfect of machines. M. Verdet however
makes no allowance for dissociation, but assumes that the
whole of the gas is consumed before the piston has moved
appreciably ; the pressure therefore reaches 14·56 atmospheres, and assuming the cylinder to be large enough to
allow the pressure to sink to one atmosphere, the products of combustion are discharged into the air completely burnt, but at 3674° F.

If, however, adopting the results of Bunsen's experiments described in the last chapter, we assume with M.
Verdet that the maximum temperature is reached while
the volume of the gases remains constant, and further that
as the piston moves forward the additional quantity of
gas consumed is just sufficient to prevent any fall in the
temperature, we arrive at very different results. The
initial pressure is only 7·7 atmospheres, and when the
gases are discharged at atmospheric pressure, not only is
the temperature practically the same as that deduced in
the last case (3627° F. instead of 3674°), but a quarter of
the carbonic oxide remains unburnt. The proportion of
the heat which the complete combustion of the fuel would
produce which is converted into work is only ·247 instead
of ·408.

This corrected result must not however be taken as
correct in itself, but only as proving the importance of not
neglecting the phenomena of dissociation if such calculations are attempted. The degree of efficiency which it
indicates is far greater than that attained in any real gas
engine. Thus Lenoir's was subjected to an elaborate series
of experiments by M. Tresca, who came to the conclusion
that only four-hundredths of the heat was utilised. In
a more modern form, Otto and Langen's, about one-tenth
of the heat is converted into work. But although this

result is as good as that attained in any but the very best steam engines, the expense of gas as compared with coal puts all gas engines at a serious disadvantage from an economical point of view.

The electro-magnetic engine is however that from which the greatest things are generally expected, and one not uncommonly hears in general conversation the most extravagant suppositions put forward as to the feats it will probably perform in the future. The fallacy which underlies many of these speculations has been frequently pointed out, but a comparison of the steam engine with other motors would be incomplete without some reference to it. Before discussing the relative advantages of engines driven by steam and electricity, let us, to fix our ideas, consider the construction of some one of the latter class. A galvanic current may be used to set bodies in motion in several different ways, but the fact which has been turned to account in most electric motors is that if a current be passed through a wire which is wound round a bar of soft iron, the iron will be converted into a magnet while the current flows, but will lose its magnetic properties as soon as the current ceases to circulate. Such a bar of iron is called an electro-magnet, and the rapidity with which it can be alternately magnetised and demagnetised makes it a useful instrument for imparting motion to other masses of iron.

A well-known form of electric engine is represented in Fig. 56. It consists of a wheel, to the circumference of which eight bars of soft iron are attached, at equal distances. Four electro-magnets are placed close to the wheel, and matters are so arranged that the current circulates in the wire wrapped round each only when one of the bars of iron is approaching and near to it, and is interrupted as the iron passes it. The bars of iron

are thus always pulled by the attraction of the temporary
magnets in the direction in which the wheel is rotating,
and never in the direction which would retard its motion.
A continuous and rapid revolution can thus be maintained. The passage of the current is regulated by a
cog-wheel, the projections on which act upon rollers attached to metal springs. These when pressed by the
cogs touch a fixed metallic button and complete the circuit

FIG. 56.—Electric Engine.

through which the current passes, but when the pressure
is removed they regain their original position and the
current ceases to flow. The electro-magnets are thus
alternately active and inactive at the proper times.

In order to understand fully the source of the energy
of this or any other form of electric engine, it will be
necessary to dwell upon the circumstances under which a
galvanic current is produced. If some granulated zinc
be dropped into dilute sulphuric acid a copious evolution

of gas takes place, and the zinc and acid are warmed. The zinc displaces the hydrogen, and unites with the sulphur and oxygen of the sulphuric acid, sulphate of zinc is formed, and the hydrogen as it is liberated rises to the surface of the liquid. The heat produced is due to the molecular motion set up during the substitution of the zinc for hydrogen, and amounts to 1018·91 units for each pound of zinc dissolved.

If the zinc is either perfectly pure, or is amalgated by moistening it with acid and dipping it in mercury, no chemical action takes place, and it is not attacked by the acid. If however a plate of platinum be also immersed in the acid, and the two metal plates (which should not touch in the liquid) are connected outside the liquid by a metallic wire, the following phenomena are observed :

(1). Though no platinum is dissolved, hydrogen is evolved at the surface of the platinum.

(2). The zinc is dissolved, forming as before sulphate of zinc, but no hydrogen is evolved from the surface of the zinc.

(3). An electric current circulates, passing through the wire from the platinum to the zinc and through the liquid from the zinc to the platinum.

(4). Heat is produced both in the liquid and the wire.

The first of these phenomena we need not now discuss ; suffice it to say that, as before, the zinc displaces the hydrogen in the sulphuric acid, though the hydrogen first makes its appearance at the platinum plate. With respect to the fourth, experiment readily proves that the amount of heat produced in the wire is very small if it is short, thick, and a good conductor of electricity.

Now we know that the electric current thus generated is capable of doing work, as it may be made to drive an engine, and the question arises—What relation, if any, exists between the work which the current can do, and the heat which appears in the circuit.

This question has been answered by M. Favre. He employed the apparatus represented in Fig. 57. It consists essentially of a large reservoir of mercury, into which six tubes, closed at the lower end (only two of which are shown in the figure), penetrated. The amount of heat produced by any operation conducted in one of

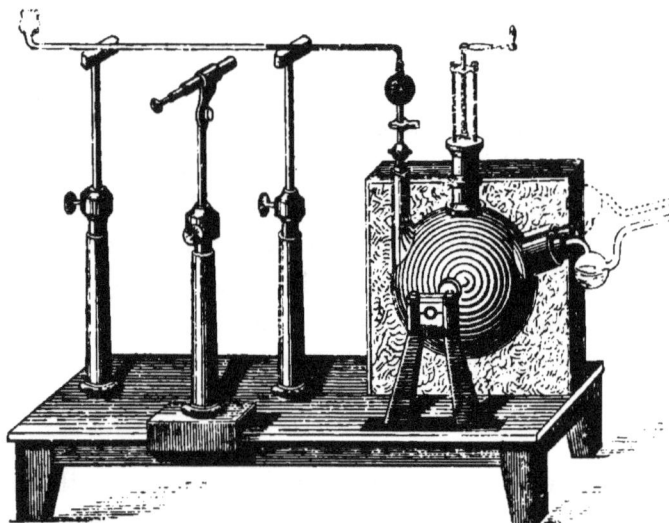

FIG. 57.—Favre and Silbermann's Calorimeter.

these tubes could be determined from the expansion of the mercury in a horizontal glass tube.

Favre first dissolved some zinc in sulphuric acid, and found, as above stated, that the solution of each pound of zinc was attended by the evolution of 1018·91 units of heat. He then placed in five of the tubes small galvanic cells, each of which consisted of plates of zinc and platinum immersed in acid, and connected them with each other by short, thick, copper wires. As under these circumstances nearly all the heat was produced in the cells, and only a very minute quantity in the wires (which were necessarily outside the apparatus), it could be measured and the amount was found to be practically the same as in the last case, viz. 1018·48 units per pound of zinc dissolved.

A small magneto-electric machine was then placed in the sixth tube. It was connected with the cells; but was not allowed to move when the current passed. The heat evolved was 1018·09 units per pound of zinc.

The arrangements being in other respects the same, the machine was next allowed to move, but did no work. The energy of the current was thus spent only in over-coming the friction of the machine, and in heating its various parts. This heat was given up to the mercury as before, and the total amount registered was nearly the same as in previous experiments, viz. 1017·55 units. This proved that the solution of the zinc must have been attended with the production of less heat than when the machine was at rest. The heat thus lost was the source of the work necessary to maintain the motion of the machine. This work was however the exact equivalent of the heat produced in the various parts of the engine, and as this too passed into the heat-measuring part of the instrument, the total quantity of heat registered was the same as before, though it had been produced in a some-what different manner.

In a last experiment the machine was made to do work in raising a weight, and now for the first time a noticeable difference was observed in the quantity of heat produced. The mean of all the previous experiments gave 1018·26 units of heat per pound of zinc dissolved, but the number now fell to 1002·12, and for each missing unit 806 foot-pounds of work had been done.

The outcome of these experiments is that any work done by the electric engine is done at the expense of the heat produced by the solution of the zinc, and it follows that if the whole of this heat could be utilised, about 800,000 units of work would be done for each pound of zinc consumed.

Heat is therefore the source of the energy of a machine driven by an electric current, as well as of that of the steam engine itself. In the latter we employ the heat generated by the union of carbon and hydrogen with oxygen, in the former we use that which is produced when zinc takes the place of hydrogen in sulphuric acid. In each a certain quantity of heat is placed at our disposal, and in neither can we get more than 772 foot-pounds of work for each unit of heat. The electric engine is, however, the more efficient instrument for the conversion of heat into work, though in it as in the steam engine we are practically unable to utilise more than a fraction of the heat employed. The causes of this imperfection are different in the two cases. When an electric current is used to make an electro-magnetic machine perform useful work, the disappearance of heat is not the only phenomenon observed. The current which drives the machine is itself weakened, and the more perfect the machine is, the larger the ratio of heat utilised to heat supplied, the greater is the decrease in the strength of the driving current. Thus every improvement in the machine, though enabling us to make a better use of the heat furnished by the battery, diminishes the quantity of heat produced in a given time. At first the gain will more than balance the loss, the increased efficiency will compensate the failing intensity of the current. At last however a point will be reached when the very perfection of the machine will itself be a disadvantage. Every further advance will give us a more efficient but weaker engine. The heat given out by each pound of zinc consumed will be more completely converted into work, but the zinc will be

longer and longer in consuming, the work will take longer
and longer to perform. A perfect electric engine would
theoretically transform all the heat supplied to it into
work, but in so doing it would entirely destroy the current
which animated it, and thus itself cut off the supplies of
heat upon which its power depends.

Although therefore it would be as impossible with the
magneto-electric machine as with the steam engine to
convert all the heat supplied into work, it is nevertheless
possible to work it at a much higher rate of efficiency.
The maximum amount of work which the machine is
capable of producing in a given time is obtained when the
efficiency is one-half, and this result has been practically
attained. A part of the work produced is lost in over-
coming the friction of the machine, and there are other
special sources of loss, such as sparks, which do not occur
in the steam engine.

The great drawback to the electric engine is the cost of
the substances employed to produce the current. That
machine will evidently be the best for practical purposes
which supplies us with each foot-pound of work at the
cheapest rate, and in this respect electricity cannot compete
with steam. The price of zinc is from thirty to forty
times greater than that of coal. A pound of coal will in
a good steam engine give us 1,000,000 foot-pounds of
work, a pound of zinc in a good electric engine will give
only 400,000. To obtain the same amount of work in the
two cases we should therefore have to use $2\frac{1}{2}$ lbs. of zinc,
costing (in round numbers) 90 times as much as a pound
of coal. Of course in a strict comparison of the relative
cost of working the two engines other elements must be
considered, but these I need not here discuss. It may be
stated generally that the cost of the electric engine
is fifty to sixty times that of the steam engine for

equal quantities of work. In other words the amount of work which the steam engine performs for a penny is only obtained from the electric engine by an outlay of about five shillings. It is evident therefore that unless some much cheaper method of generating electricity is discovered, the electric engine will never be employed in cases where the steam engine is equally available.

The last class of machines with which we shall compare the steam engine are of Nature's own contriving, viz., the bodies of animals. These derive their power of performing work from the heat produced by the slow combustion of their food, and may therefore for our present purpose be regarded merely as heat engines. It might at first sight appear that these were not subject to the laws we have been discussing inasmuch as active labour induces a rise in the temperature of the body. It might thus seem as if work was not done at the expense of the heat produced in the system, but that living beings possessed the peculiar power of producing work and heat together without the expenditure of either. It must however be remembered that labour increases the rapidity of respiration. The internal fire, which burns low when the animal is at rest, is fanned, and of the additional heat thus produced some is spent in warming the body, some disappears in the accomplishment of work. Thus a person ascending a hill consumes more oxygen in a given time than if he were at rest, but he gets less heat from each pound of oxygen consumed.

The production of heat in the muscles of animals takes place also under somewhat peculiar circumstances. Thus when a muscle contracts it becomes heated, but the elevation of temperature is greater when it is by its contraction made to lift a weight, than when it is unloaded and performs no work. This result, though at first sight paradoxical, is not out of harmony with the theory that the performance of work requires the expenditure of energy. But little is known of the chemical or electrical phenomena of which we do know that a muscle is the seat, and we can easily suppose that the changes which occur in stretched contracting muscle, result not only in the production of work, but also in that of more heat than if the muscle were free. If the same internal changes could be brought about while the muscle did no external work more heat would be produced than is actually observed.

If water between 32° and 39°.2 F. be compressed, it is cooled; while if it be above 39°.2 F., compression produces the evolution of heat. The fact that equal contractions in two muscles, in the different molecular states resulting from a slight difference of tension, are attended with unequal thermal effects is no more incomprehensible than the fact that if two masses of water in the different molecular states resulting from a slight difference of temperature are equally compressed, the one will be cooled and the other heated.

The efficiency of the animal economy may be measured by the relation between the total quantity of heat which could be produced by the combustion of the food which the animal consumes in a given time, and the heat which would be necessary to do the amount of work which it is capable of performing in that time.

Several attempts have been made to investigate this relation. Dr. Joule determined the quantity of heat produced by burning a known weight of a fuel composed of equal parts of corn and hay in oxygen. Assuming that a horse requires 12 lbs. of hay and 12 lbs. of corn per diem, he found that the heat produced by the combustion of its food would be sufficient to do 88,500,000 foot-pounds of work. A horse however can at the most do about 24,000,000 foot-pounds of work per day, so that only about $\frac{3}{11}$ of the heat produced by the combustion of its food can be utilised. The efficiency of the horse regarded as a heat engine is therefore about three-elevenths.

From this point of view a man is probably an inferior machine to a horse. Curiously enough a healthy adult at rest produces in an hour just enough heat to raise his body through the height which he can conveniently climb in an hour on a hill offering no special obstacles to his progress. When actually climbing however he breathes five times as much air as when at rest. In order therefore to raise his body through a certain height, he consumes fuel enough to do five times the amount of work he actually performs. In other words his efficiency as a machine is one-fifth.

These calculations are however of a very rough character, and can only be considered as first approximations to the truth. They may suffice to show that the loss of heat even in the animal economy is very large; although, regarded as a heat engine an animal is probably about twice as efficient as the best steam engine.

The subjoined table gives in the second column the percentage of heat which the various machines named transform into work under favourable conditions, and in the third the relative cost of the fuel capable of producing equal quantities of work in each.

	Percentage of heat utilised	Relative cost of fuel for equal quantities of work
Electric Engine	50	55·00
Air Engine	13	0·84
Steam Engine	11	1·00
Gas Engine	10	6·00

It is evident from the above discussion that the steam

engine is not likely soon to lose the pre-eminence which it at present enjoys. The electric engine and the gas engine are convenient, both because they can be set in motion without delay and because the expenses connected with them are incurred only when they are actually working. They are compact in form, and are well suited for operations in which a small amount of power at irregular intervals only is needed. The cost of the fuel is however in both cases too great to admit of their employment on a large scale. The air engine on the other hand, though making a better use than the steam engine of the fuel with which it is supplied, is discarded on account of its bulk and the other disadvantages enumerated above.

For the moment however our interest in these machines is limited to their relations to coal, and it does not appear that any changes in their relative merits which future inventions and improvements may possibly bring about are likely to diminish the paramount importance of mineral fuel. In the steam and air engines it is employed directly, from it the gas required for the gas-engine is extracted, and by it the zinc or other metals used in the galvanic cell are smelted. In all these cases therefore the source of power is the same, though the machinery for transforming its energy into useful work is widely different, and should coal fail our descendants as wood fuel threatened to fail our ancestors some three centuries ago, the wind, the waterfall, and the tide will probably be the only stores of energy on a large scale which will be left at their disposal.

CHAPTER IX.

THE COAL QUESTION.

THE phrase which heads this chapter is now by common
consent confined to that one among many possible questions
regarding coal which deals with the conditions of its production and consumption as an article of wealth : more
especially in our own country, where, more immediately
than in any other, these conditions are of pressing and
special importance. The causes of the peculiar interest
attaching to the question of our coal supplies and how
we use them, are two-fold. In the first place, our stores,
great though they be in comparison with those of many
other important states, are nevertheless limited ; and the
limitations are definite and calculable : moreover, while
these stores are great in this aspect, they are not immeasurably great compared with the drain which is year by
year being increasingly put upon them : so that as a matter
of fact, whatever differences may appear as to the time
and manner of the future exhaustion of our coal-fields,
there is, we shall find, no material controversy on the bare
proposition that exhaustion is ultimately inevitable.

In the second place, the preceding chapters will be
enough to show that to us, more than to any other nation
on the globe, dearth of coal means material ruin. The
enormous changes effected both here and elsewhere during
the last hundred years by the aid of British coal have
been described again and again. We are perfectly familiar

with such facts as that the population of this country has increased five-fold during that time : that the exports and imports have increased considerably over ten-fold : that in the interval portions of the globe amounting in area to something over a hundred times the area of the British Isles have been largely colonised and settled by British subjects, and are now the seats of ordered industry and intelligence comparable with those of the mother country herself: and so with endless other facts, all tending to show that a development of power and an increase of wealth have been witnessed in the past century, with which nothing whatever in the previous history of the world can be compared.   Two main causes may be adduced for these wonderful phenomena,—the growth of our coal-driven industries and the liberation of labour ; the period begins with the invention of the steam engine : it also begins with the French revolution. It is outside our pur-pose to discuss the latter of these influences, one which of course is by no means entirely or mainly British : but it is questionable how far the growth of popular freedom could have reached, consistently with the preservation of culture and the liberal arts, had it not been that as the chains fell from the labouring classes of Europe, a new bondsman more powerful far· in some respects than the other, was found in coal.   When we consider that in the coal raised in Great Britain alone in 1876 an amount of energy was contained equal to the labour of more than 3,000,000,000 adult slaves labouring daily throughout the year we may well understand how it is that human slavery has so utterly died out in all civilised, that is, all coal-using, countries since 1778.

But however this may be, there is no question that for good or evil the prosperity of Great Britain, and with her that of the world of which she is the commercial centre,

have become bound up with the uses she has learned to
make of the coal she possesses so abundantly, and Prof.
Jevons was therefore justified in placing as the title of
his work on this subject the following words: 'The Coal
Question; an Inquiry concerning the progress of the
nation, and the probable exhaustion of our Coal-mines.'
Whether we are to suppose that there can be no progress,
no prosperity for this country when the day comes when
her coal is exhausted will be discussed towards the close of
these chapters: for the present it is sufficient to say, that
the Britain with which we are familiar as the factory, the
mart, the exchange of the world: with its crowded populations, its vast wealth and splendid material achievements,
the very paradise of self-satisfied statistics, cannot possibly continue to exist when the coal which is the main-
spring of its energies has disappeared; and that there can
hardly be a doubt that the period of transition from
our present position to that which our resources other
than coal could support, will be one of social and political
trial such as modern statesmanship never yet has had to
face. We have accumulated and are accumulating immense debts, great populations, and wide-spread traditions
of social comfort and political power among those populations, all which will make the turn of this mighty tide a
grave and ominous time.

A very sufficient proof of our present absolute depend-
ence on Coal was afforded us by the 'Coal-famine,' as with
some exaggeration it was called, of 1872–73. The con-
fusion and distress then occasioned to us all by a deficit
amounting in the whole to not more probably than one
per cent. of the world's supplies of coal, or two per cent.
of our own, and the subsequent prolonged depression and
even collapse for which the high prices resulting from
that deficit are largely responsible, are only a premonition

of what would ensue, were such a deficit to become chronic, and a real coal-famine came upon us.

What the Coal-question then amounts to is simply this: Is there any probability of such a famine in Great Britain within a period to which our interest can extend, say within a century or two: and what, if any, are the signs of its approach?

As a mere matter of arithmetic, supposing the facts are within our reach, the question might seem, at a first glance, a tolerably simple one. Given the quantity of coals still unworked, and the amount by which we annually reduce that quantity, we shall simply have to divide the first by the second to obtain the number of years for which our total will last. But there are few that will imagine, even for a moment, that the matter is at all so simple as this. We have, doubtless, as will be shown later on, full and satisfactory information both as to the total quantity and as to the annual output; but with this information before us, our difficulties are but begun. For apart from all questions, and they are not few, as to the proportion of our total amount which will ultimately prove commercially or even physically attainable, one glance at the output will show that the divisor of our little sum is not so easy to determine: year by year we shall find that divisor growing, without a sign of pause up to the present moment: under such conditions the arithmetical problem at once becomes a question of future probabilities: and this, a speculation the most complex and difficult conceivable.

While, therefore, it is proposed to begin with a statement of the known facts as to, *first*, our total quantity, and *second*, our annual output, and to show certain numerical relations between these for the purpose of illustration, the reader is requested to observe that only

in the *general* results of the whole discussion, and not in
such numerical calculations, is any approach to a theory
of the probable duration of our coal-fields contained.
Abundant reason exists for such a caution, for there is the
greatest temptation to snatch in subjects of this kind at
any apparently definite statement apart from its context,
and the figures thus 'untimely ripp'd' are forthwith
bandied about as the ' theory of such and such a writer,'
when it is extremely probable that they represent no more
than a supposed theory mentioned only to be rejected.

With this caution then let us proceed to deal with
what may be called the historical and generally uncon-
troverted section of our inquiry.

Although, as has been shown in a previous chapter,
some of our great coal-fields had been worked to some
extent for several hundreds of years, prior to 1800 no
approach even to a theory either of the quantity of coal
we possessed, or as to the rate at which it was being consumed, seems to have been attempted. There was in fact
neither the inclination nor the means to do so. Those
who were in any degree intelligent agents in the development of our coal-production, had a task amply sufficient
for their powers, in the discovery of new or better ways
of getting the coal, or of utilising it. Such men as Darby
and Watt and Murdoch were gifted with too practical a
genius to think of a question apparently so remote. On
the other hand, the general notion with respect to our
coals, as evidenced in the daily conduct of those who
handled them, seems to have been that the supply was so
absolutely inexhaustible that the most enormous and
obvious waste was a matter of no moment. Thus we are
told[1] that in the great Staffordshire Coal-field, the splendid
' Ten-yard Seam ' as it is called, was so recklessly worked,

[1] *History of Fossil Fuel*, p. 433. Whittaker; 1841.

that at least two-thirds of the entire seam were lost in
mining. At present the loss in mining under similar
conditions probably does not exceed one-fourth. The same
writer observes that 'a still more lamentable waste of
excellent coal took place in the South Welsh and more
particularly in the Northern Collieries, at the pit mouth,
in consequence of the practice of screening. This was
done to meet the taste for round coals so generally
prevalent in the metropolis, and also to meet the circumstances of a demand which, before the trade imposts were
reduced, and weight substituted for measure, required the
coals to be shipped of a large size.' The waste thus
caused was estimated by Mr. Buddle in 1829 as amounting
to from one-fourth to one-third of the total output, nearly
the whole of which was burned in heaps at the pit-mouth ;
and it is said that in some cases from 100 to 200 tons
were destroyed in this way at a single colliery *per diem.*[1]
What is perhaps more surprising is, that all this waste
was absolutely without result even for the purpose for which
it was intended. Evidence was given before a Committee
of the House of Commons which showed that it was the
universal practice on board ship to break up the round
coals that had been obtained in so costly a fashion, in
order to increase their apparent bulk when measured in
the port of London, so that according to one witness ' they
usually reached port inferior in point of size to what they
would have been if put on board unscreened.'[2]

These after all are but types of the recklessness with
which our coals were squandered a hundred years ago :
and we should, therefore, hardly expect to find any great
anxiety then as to the statistics of the matter, when the
coal itself was so little valued. It is true there is evidence

[1] *History of Fossil Fuel,* p. 434.
[2] *Ibid.* p. 382.

of immense wastefulness even now, but where such wastefulness exists, the same spirit is shown in the ignoring of
statistics and their teachings, as was formerly shown in
the fact that such statistics did not exist.

We may, moreover, say with truth, that before the
beginning of this century the value of statistics as means
of accurate knowledge or as guides to conduct, social or
political, was little understood. The first census of the
population of England and Wales was taken in 1801,
and the first of the population of Ireland only in 1821;
and the traditions of that earlier period of blind and easy
confidence in a future for which no reasonable provision
is made, are by no means yet extinct.

It was, however, impossible, even had there been the
desire, to obtain the most approximate estimates of the
amount of coal existing in these islands, or consequently
of its probable duration, till the principles and methods
of geological inquiry had been developed, and some tolerably complete synopsis of the geological history of the
country in accordance with those principles had been
obtained. It is only within the last quarter of a century
that this has been done, so that none but the most recent
writers on the subject have been in possession of one of
the two indispensable elements of any calculation with
respect to the duration of our coal-fields.

Parallel with this very gradual approach to accurate
knowledge on the question of our total stores of coal,
attempts were made from time to time also to obtain an
estimate of the output. The estimated amounts for certain
dates prior to 1800 have already been given (page 234):
there are here subjoined certain estimates for dates subsequent to that year, with their respective authors.[1]

---

[1] See Royal Commission Report on Coal, iii. 61, 62.

Year.	Estimated production.	Author of estimate.
1816	15,634,729 tons.	Report of deputation from the Wear, 1816.
ditto	27,020,115 „	'Statistics and Calculations' of Saml. Salt.
1829	16,034,799 „	Hugh Taylor.
1839	31,024,417 „	J. R. M'Culloch.
ditto	36,000,000 „	H. T. de la Beche.
1845	34,600,000 „	J. R. M'Culloch.
1846	36,400,000 „	J. Emerson Tennant.
1852	34,000,000 „	Braithwaite Poole.
1851, 1852, 1853 each	54,000,000 „	Jos. Dickinson.
1854	56,550,000 „	T. Y. Hall.
ditto	52,000,000 „	Ditto (quoting another authority).
1855	58,200,000 „	J. R. M'Culloch.

It will be observed that the discrepancies between the estimates of different authorities for identical or nearly identical years are in some cases very great—amounting in 1816 to a difference of 80 per cent., and in 1852 to nearly 60 per cent. No such discrepancies appeared in the estimates previously given for years prior to 1800: but the reason is that these are *recent* estimates made on one system by one inquirer; those are *contemporary* estimates, by different authors, with varying means of information, and equally various powers of judgment.

For the earlier estimates, the chief source of information was the records of a curious institution called the 'Limitation of the Vend,' by means of which the Coal-owners of the Tyne, Wear, and Tees systematically limited their own production year by year, with a view to their own profit. The history of the institution no further concerns us here, than that we possess the records of that production for certain years during last century; and it has been estimated that on the whole, this recorded production probably constituted one-fourth of the total coal-production of Britain. The estimates of the years given are therefore the results simply of a quadruplication of

the recorded outputs of the Northern Coal-fields. Whatever may be the value of such a mode of calculation for
other purposes, it is obvious that no deductions as to a
general law of increase in the output, applicable to our
own times, can safely be drawn from such questionable
premisses.

For the period between 1800 and 1854, the discrepancies
already referred to show that our materials are equally
unsatisfactory: and we are the more confirmed in this
opinion when we compare the latest and highest of those
estimates, with the actual output for the years 1854, 1855,
ascertained in a way to be described immediately. That
output was for 1854 *at least* 64,661,401 tons; for 1855,
64,307,459 ; in each case exceeding the estimates by from
6,000,000 to 12,000,000 tons.

For the more accurate returns from 1854 onwards to
the present day we are indebted to the labours of Mr.
Robert Hunt, F.R.S., Keeper of the Mining Records
Office, who under Government sanction, with the assist-
ance of the Mining Inspectors, and aided by the co-operation of the Colliery owners, the Railway Companies, and
others, has been enabled to obtain year by year a statistical
resumé of our coal-production, which if not absolutely
accurate, is as nearly so as are most statistics in this imperfectly tabulated world.

Complete knowledge of the total *quantity* of coal came
somewhat later. The subject became one of public interest
in 1860, the year of the commercial treaty with France,
in consequence of the discussions which arose on the
subject of our exportation of Coal to that country. Mr.
Edward Hull, F.R.S., Director of the Geological Survey
of Ireland, thereupon undertook the task of forming a
careful estimate of the total quantity of coal existing in the
known coal-fields of this country, and published his results

in a well-known work, 'The Coal-fields of Great Britain.'
By 1863 therefore something like satisfactory materials
existed for an estimate of the relation between our total
stores and our output, and accordingly Sir William
Armstrong in his Presidential address to the British
Association at Newcastle in that year, directed the attention of his audience to the alarming growth of the output
relative to this total. With the materials before him he
calculated that, assuming an arithmetical increase in output similar to that obtained by the average observations
of the eight years preceding, our total stores to a depth of
4,000 feet would be exhausted in 212 years from the year
1863; and he pointed out that (according to principles
which will be discussed later on) actual exhaustion must
take place at a very much earlier period.

It is probable that Sir William Armstrong's address
directed Prof. Jevons' attention more particularly to the
subject : at all events his celebrated work already referred
to, appeared in 1865. Its main purport was to show that
Sir William Armstrong had erroneously interpreted the
rate of growth in the output of coal, and that as a matter
of fact there was proof that the rate of growth was a
geometrical one ; and that consequently, on the assumption
of its uninterrupted continuance, our coal would be exhausted in 110 years from 1861. The details of his arguments will be referred to later on; what is desired here is
to show merely the historical development of the inquiry.

Prof. Jevons' work, from the remarkable thorough-
ness and lucidity of its argument, and from the alarming character of its conclusion, caused something like
a general panic; and one result, the most important
for our present purpose, was the appointment of a Royal
Commission in 1866, ' To investigate the probable quantity of coal contained in the coal-fields of the United

Kingdom, and to report on the quantity of such coal
which may be reasonably expected to be available for
use; Whether it is probable that coal exists at workable
depths under the Permian, New Red Sandstone, and other
superincumbent strata; To inquire as to the quantity of
coal at present consumed in the various branches of
manufacture, for steam navigation, and for domestic purposes, as well as the quantity exported, and how far, and
to what extent, such consumption and export may be
expected to increase; And whether there is reason to
believe that coal is wasted by bad working, or by care-
lessness, or neglect of proper appliances for its economical
consumption.' Fifteen Commissioners were appointed,
the Duke of Argyll being chairman, and among the rest
were the names of Sir Roderick Murchison, Sir William
Armstrong, and a number of the most eminent practical
engineers and geologists in Great Britain.

Their inquiry was concluded and their report published
in 1871, and the very complete and elaborate information
therein contained, together with the subsequent volumes
of the 'Mineral Statistics' constitute the accepted and
undebatable ground of fact from which every theory on
the subject of our coal production must start, and to which
every such theory must be referred. It might perhaps
suffice to give a summary of the conclusions of the Commission as to the total quantity of coal in the kingdom,
'which may be reasonably expected to be available for
use' were it not that on several points these conclusions
have been disputed, on grounds which are not unfrequently
referred to as if they were new elements in the question,
whereas they are for the most part merely reproductions
of alleged facts or opinions, which were submitted to the
Commission and, having been considered on their merits,
were rejected or disproved.

The chief points which have thus been re-opened are
—the probable limits of depth in working—the limits of
coal-bearing areas—and the thickness of workable seams.

The generally accepted limit of depth in working is
4,000 feet from the surface. No such depth has ever yet
been attained in mining; only a very few mines in this
country in fact greatly exceed half that depth, although it
has been approached in Belgium, where one mine, now
abandoned, that of des Viviers, at Gilly, near Charleroi
was for some time worked at a depth exceeding 3,400 feet.
The reason of this purely theoretical limitation to 4,000
feet is the observed increase of temperature as we descend
into the earth. Over England the temperature of the
earth is uniformly 50° Fahr. at a depth of 50 feet on the
average from the surface, and the evidence given before
the commission went to show that the temperature of the
strata increased in general about 1° Fahr. for every 60
feet of depth. A temperature of 98°, or blood-heat is thus
reached in the strata at a depth of 3,000 feet. Under the
long-wall system of working, a difference of about 7°
appears to exist between the temperature of the air and
that of the strata : the air therefore would reach blood-
heat at a depth of 3,420 feet; and a further addition up
to 4,000 feet was made to cover possible expedients for
artificially reducing the temperature.

The limit of 98°, or blood-heat was, on the evidence,
adjudged by the Commissioners to be the highest at which
men could work efficiently in a humid atmosphere such
as is usually found in coal-mines; 4,000 feet therefore
they concluded was the limit of possible working, and
their calculations of the amount of available coal were
taken on that footing.

It has been since maintained that the ratio of increase
in temperature as we descend is a diminishing one, and

that the limit of working may therefore have to be extended. So far as actual experiment seems to support such a statement, the whole of the observations were before the Commission, and were deemed inconclusive : since that time it remains entirely unconfirmed. It is indeed true that eminent physicists, such as Sir William Thomson, on theoretical grounds admit the possibility or even probability that such is the case, but no *sensible* diminution of the ratio is considered probable at a less depth than 100,000 feet.

The limit, however, it must be admitted, has been so nearly approached without special inconvenience, when efficient ventilation and suitable methods of working have been adopted, that it seems highly probable that in time certain valuable seams in this and other countries may be worked at a depth beyond the limit assigned by the Commissioners. But it is in the last degree improbable that this will be the case under any other than exceptionally favourable circumstances, and if we take into account the very high probability that many deep seams will never be worked even up to that limit, from *economical* considerations, the principle adopted of counting in *all* seams to the depth of 4,000 feet, whatever the local conditions or economical difficulties of their working, will not seem to have erred on the side of timidity.

As to the limits of coal-bearing areas, it will be observed that it was an instruction to the Commission to inquire ' Whether it is probable that coal exists at workable depths under the Permian, New Red Sandstone, and other superincumbent strata,' and in the result it will be found on inspecting the table of the British coal-fields (p. 308) that a very large addition to the estimated total was made on account of coal-bearing areas concealed under these more recent formations. The most important of these concealed coal-fields are those bordering on the Yorkshire and

Derbyshire, and the South Staffordshire coal-fields; and in both these districts several successful attempts have been made to reach the coal-seams through the overlying strata, as near Doncaster, in the vicinity of Birmingham, and elsewhere. These successes, however, so far from being regarded, as they ought to be, merely as confirmations of the soundness and accuracy of the Commissioners' conclusions, are frequently referred to in popular discussions on the subject, as if they cast doubt on the completeness of their results, as if, in fact, they were new and unexpected discoveries. Until coal is discovered beyond the calculated limits adopted in the Commissioners' report, no addition, of course, can be made to their calculated amounts.

Controversy arose, however, within the Commission itself on the question of the theoretical probability of a great coal-bearing area existing within workable depths in the South of England. From a consideration of certain analogies between the Belgian and the Bath and Bristol coal-fields, and for other reasons, it was maintained by Mr. Godwin-Austen, and admitted by Mr. Prestwich, one of the Commissioners, ' that the coal measures which thin out under the chalk near Therouanne, probably set in again near Calais, and are prolonged in the Thames Valley, parallel with the North Downs, and continuing thence under the valley of the Kennet, extend to the Bath and Bristol coal area.' This view however was strongly controverted by Sir Roderick Murchison, who contended ' that in consequence of the extension of Silurian and Cambrian rocks beneath the secondary strata of the S. E. of England, and of the great amount of denudation which the Carboniferous rocks had undergone over the area of the South of England previous to the deposition of the secondary formations, little coal could be expected to remain under the Cretaceous rocks.' On the whole the Commissioners decided to leave the question unsettled, and no estimates were made as to the quantity of coal (if any) in this unexplored area.

Should the attempt ever be successfully made to reach a workable area of coal in this district, the calculations of the Commissioners would of course have to be materially enlarged : but the only boring yet attempted in the Weald, although continued down to a depth of 1762 feet, showed no trace of coal.

With respect to the limit of thickness for workable seams of coal, the instruction to the Commissioners to whom the several districts were assigned was to exclude from their returns all beds of coal of less than one foot in thickness. With regard to this instruction it has been urged by many competent persons that the limit therein fixed was too low : that even at the more moderate depths it is only under exceptional circumstances that seams of less than two feet in thickness can be worked at a profit, as they are for example at Low Moor near Leeds, where the seams are of exceptional purity ; and that as still greater depths were approached the commercial impossibility of such workings would become all the greater.

It is impossible to deny that this objection has some apparent force. In the examination of Mr. Greenwell[1] it was shown, that in the case of certain very thin seams which were worked in Somersetshire on the long-wall system, one seam being as thin as 14 inches, another $16\frac{1}{2}$ inches, worked some of them at a depth of 1260 feet, the circumstances were peculiarly favourable. Under the coal there is a sort of soft underclay very easily worked, so that practically the whole of the coal is extracted. In answer to a question whether seams of such a thickness

[1] Royal Commission Report, ii. 337.

could generally speaking, be worked at a profit, the answer
of the witness was, that it entirely depended upon the selling
price of the coal : that the Somersetshire coal-field was a
long way removed from other coal-fields, making the railway charges upon other coals so high that a good price
(that is, about 10s. to 12s. per ton at the pit mouth) could
be obtained for them. There being no fire-damp it was
found quite possible for the men to work in an opening
less than 2 feet high.

I apprehend that the Commissioners placed the limit
of thickness as low as 12 inches because their inquiries
were not in that connection directed to the question what
amount of coal would ultimately be found *commercially*
workable : it was the simple physical limits which they
were chiefly regarding. Thus they say in their general
report (vol. i. p. 18) : 'Much of the coal included in the
returns could never be worked except under conditions of
scarcity and high price. A time must even be anticipated
when it will be more economical to import part of our
coal than to raise the whole of it from our residual coal-
beds; and before complete exhaustion is reached the im-
portation of coal will become the rule, and not the exception, of our practice.' From this point of view it is clear
that the Commissioners were as well justified in including
seams of this thickness, as in including seams which
approached a depth of 4,000 feet, since neither the one nor
the other would probably be worked except under identical
conditions of scarcity and high prices.

I now proceed to give in a somewhat condensed form
the returns as to the amount of coal in the exposed and
concealed areas of the several coal-fields of the United
Kingdom lying at a depth of less than 4,000 feet, after
all necessary deductions for losses in working, &c.

Name of Coal-field	Net amount of Coal in Tons in exposed Coal-fields	Ditto in concealed areas	Total
South Wales . . .	32,456,208,913	—	32,456,208,913
Dean Forest . . .	265,000,000	—	265,000,000
Bristol	4,218,970,762	400,000,000	4,618,970,762
Warwickshire . .	458,652,714	2,494,000,000	2,952,652,714
S. Staffordshire, &c..	1,906,119,768	16,189,000,000	18,095,119,768
Leicestershire . . .	836,799,734	1,000,000,000	1,836,799,734
N. Wales	2,005,000,000	—	2,005,000,000
Anglesey	5,000,000	—	5,000,000
N. Staffordshire . .	3,825,488,105	1,500,000,000	5,325,488,105
Lancashire and Cheshire.	5,546,000,000	9,085,000,000	14,631,000,000
Midland (Yorkshire, Derbyshire, and Nottinghamshire) .	18,172,071,433	23,082,000,000	41,254,071,433
Black Burton . . .	70,964,011	33,000,000	103,964,011
Northumberland and Durham . .	10,036,660,236	—	10,036,660,236
Cumberland . . .	405,203,792	1,593,000,000	1,998,203,792
Scotland	9,843,465,930	No estimate	9,843,465,930
Ireland	155,680,000	27,000,000	182,680,000
Total	90,207,285,398	56,273,000,000	146,480,285,398

An estimate was also made of the quantities of coal at depths over 4,000 feet. In the exposed coal-fields, 7,320,840,722 tons: in the concealed coal-areas, 41,144,300,400 tons: making in all a total of 48,465,141,122 tons. This amount, however, need not for our present purposes be considered.

We may take it that since these estimates were made up, about 900,000,000 of tons have been raised, so that the total available amount at present existing is about 145,500,000,000 tons.

It is of course absolutely impossible to convey any idea by description or otherwise of what these figures represent. It may be interesting however to observe, that inasmuch as a ton of coals equals a cubic yard, and an acre of coals a yard thick is therefore equivalent to 4,840 tons, a pillar of solid coal having a base an acre in extent, and containing the above total of 145,500,000,000 tons, would extend perpendicularly upwards of 17,000 miles! Or

again we may say, that its cubical contents would be represented by that of a range of hills about four miles
across at the base, and half-a-mile or 2,640 feet in average
height and extending in length about 30 miles.

If we turn now to the records of our output since
1854, we shall find that even the latest and highest recorded output, that of 1876, amounting to 133,344,766
tons, bears but a small proportion, from the simple arithmetical point of view, to this vast store of coal. The
latter in fact contains the former some 1,088 times. But
the merest glance at the returns for several successive
years will show that it would be in the last degree absurd to speak as if we had enough of coal to last at the
present rate of consumption for more than a thousand
years, since the present rate of consumption is not a fixed
but a growing rate, and there is nothing to justify the
selection of any one year's output, even the latest or highest, as in any sense representative of our future consumption. We are in fact at once met with the problem, *What
is the observed law of the growth of our output up to the
present, and What are the probabilities as to the continuance,
or modification of that observed law in the future?*

On the following page is shown a diagram intended to
represent the history of our output and export of coal
for each successive year from 1854 down to 1876, the
latest year for which complete returns are available. For
purposes of more exact reference there is also appended a
table showing the exact amounts of each, as they are given
in the 'Mineral Statistics of the United Kingdom,' to
which are added calculations of their respective yearly
increase per cent., and the yearly proportion of export to
total production.

DIAGRAM 1.

YEARS.

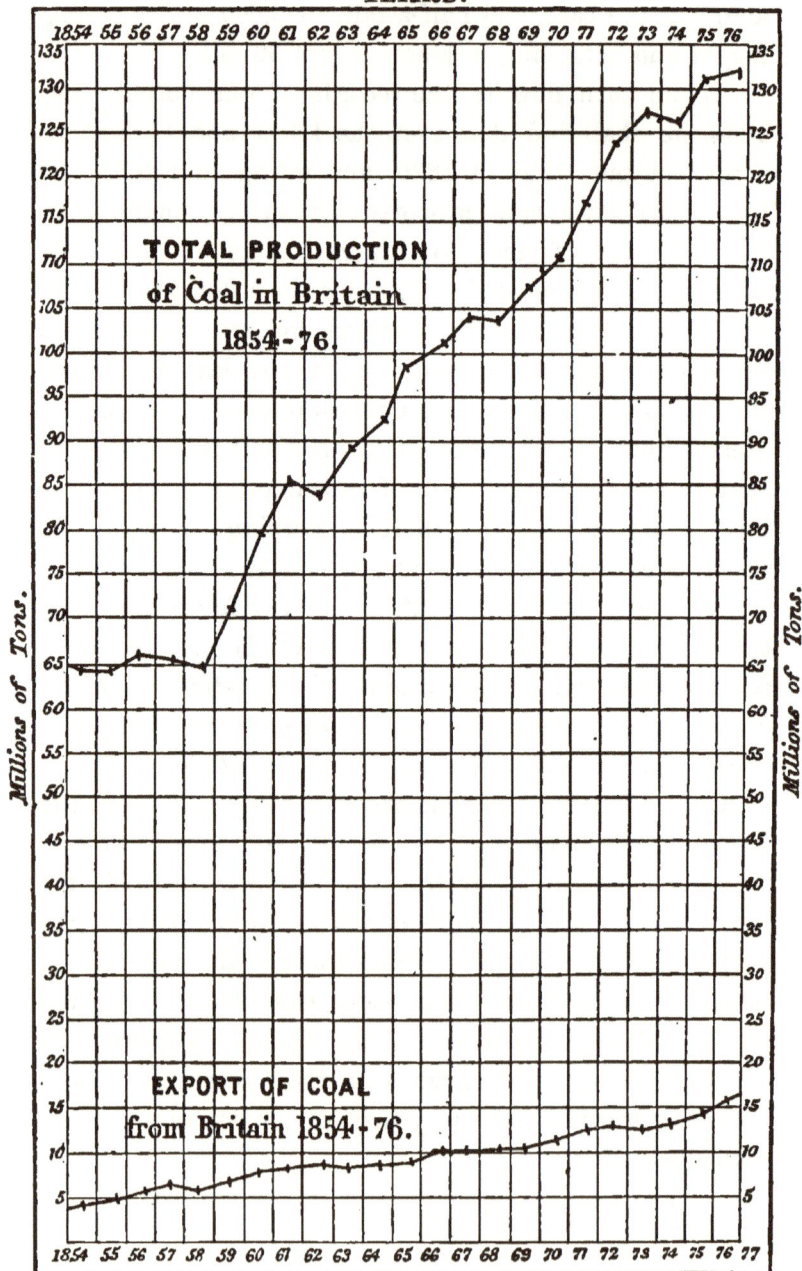

TOTAL PRODUCTION
of Coal in Britain
1854-76.

EXPORT OF COAL
from Britain 1854-76.

Year	Production	Increase or decrease per cent.	Export	Increase or decrease per cent.	Proportion to total production
1854	64,661,401		4,309,255		6·7 p.c.
1855	64,307,459	− ·5	4,976,902	+ 15·4	7·7 ,,
1856	66,508,815	+ 3·4	5,879,779	+ 18·2	8·8 ,,
1857	65,274,047	− 1·9	6,737,718	+ 14·6	10·3 ,,
1858	64,887,899	− ·6	6,529,483	− 3·2	10·6 ,,
1859	71,859,465	+ 10·7	7,006,949	+ 7·3	9·7 ,,
1860	79,923,273	+ 11·2	7,321,832	+ 4·5	9·2 ,,
1861	85,512,144	+ 6·9	7,855,115	+ 7·3	9·2 ,,
1862	83,510,838	− 2·4	8,301,852	+ 5·7	9·9 ,,
1863	88,165,465	+ 5·6	8,275,212	− ·3	9·4 ,,
1864	92,662,873	+ 5·1	8,809,908	+ 6·5	9·5 ,,
1865	98,150,587	+ 5·9	9,170,477	+ 4·1	9·3 ,,
1866	101,630,543	+ 3·5	10,137,260	+ 10·5	9·9 ,,
1867	104,509,480	+ 2·8	10,565,829	+ 4·2	10·1 ,,
1868	103,141,175	− 1·4	10,967,062	+ 3·9	10·6 ,,
1869	107,427,557	+ 4·2	10,744,945	− 2·0	10·0 ,,
1870	110,431,192	+ 2·8	11,702,649	+ 8·9	10·6 ,,
1871	117,352,028	+ 6·3	12,747,989	+ 8·9	10·9 ,,
1872	123,497,316	+ 5·2	13,198,494	+ 3·6	10·6 ,,
1873	127,016,747	+ 2·9	12,617,566	− 4·4	9·9 ,,
1874	125,067,916	− 1·5	13,908,958	+ 10·2	11·1 ,,
1875	131,867,105	+ 5·4	14,475,036	+ 4·1	10·9 ,,
1876	133,344,766	+ 1·1	16,255,839	+ 12·3	12·2 ,,

₊ For later returns, See Appendix.

The table so far as it deals with our Coal Exports, will be referred to in the following chapter.

The foregoing tables and diagram contain substantially the whole materials directly available for the purpose of our inquiry: and our references to them will necessarily be frequent. For the present we may content ourselves with a more hasty and graphic estimate of the lesson they convey as to the law of growth of our total output. Looking at the diagram one observes that roughly speaking our output has been doubled in the period there represented ; that is to say between 1854 and 1876. According to the returns, the output of 1873 nearly attained this point: but there are reasons which render it doubtful whether the returns for the earlier period, that is to say for the first six or seven years are not several millions, perhaps three or four, too low. In

the first place an error was made for several years in
estimating the real weight represented by the South
Staffordshire chaldron, resulting in an error of nearly
two millions yearly. In the next place, it is highly probable that in the earlier years, while the system was new
and strange, and the coal-masters not improbably enter-
tained some doubts as to the uses which might be made of
their returns, those returns were considerably understated.
As it is, not more than two-thirds of the coal-owners
send in complete returns; and the system of corrections
and supplementary estimates which now satisfactorily
supplies those deficiencies could not be expected to attain
perfection at once.

These considerations will affect our calculations in
other ways later on; meanwhile we may probably say with
safety that it has in fact taken twenty-two years to add
some sixty-five millions to our output. This great increase may be interpreted, however, in two ways: we may
either say that an arithmetical addition as above stated
has been made during that period, or we may say that
our output has in fact been doubled, and we may construct
hypothetically an outline of our future production on
either interpretation. The general effect of each of these
assumptions is roughly illustrated on the accompanying
diagram. (See opposite page.) Taking our total as before
at 145,500,000,000 tons, we observe that this represents
an amount roughly equal to a little over forty-seven such
blocks as that represented in the whole parallelogram of
our first diagram, in which each perpendicular column
represents 135,000,000 tons. These forty-seven blocks are
shown in our second diagram, three-fourths of a block on
the left, with darkened area, representing the actual output 1854–76. The unbroken sloping line represents the
curve of arithmetical increase for successive periods of

DIAGRAM 2.

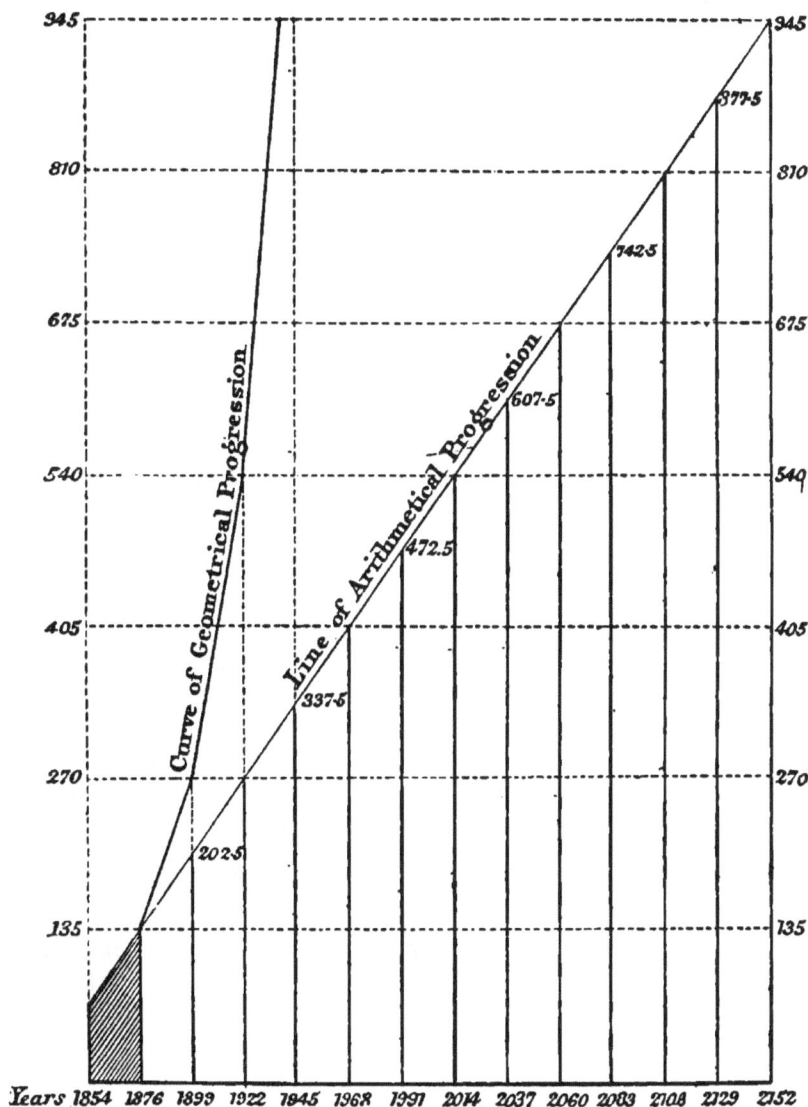

twenty-three years, and we see that the total number of blocks is exhausted at some date about 2150.

The broken curve marking a consumption of 270,000,000 in 1899; of 540,000,000 in 1922; and of 1,080,000,000 in 1945, shows some portion of the curve of geometrical progression; and it is not difficult to calculate that at such a rate of progression our whole stores would be exhausted, not in twelve periods as in the former case, but in somewhat over five; in fact in about 125 years from the present time.

Taken literally, both assumptions are obviously untenable. They imply that at some date, whether 2150 as on the former assumption, or 2000 as on the latter, an annual output which had attained the gigantic dimensions in the one case of 945,000,000 tons, in the other of over 6,000,000,000, comes suddenly to an absolute and total stop; so that in the year immediately following and for ever after not one ton of coal is ever raised more in the United Kingdom; or at all events not a ton is left anywhere to be worked at a less depth than 4,000 feet.

It is perfectly obvious that at some point in the interval a change in the present rate of growth must take place, and that influences must begin to act upon our output which shall tend to diminish the rate of increase, and so gradually divert the curve of production from an ascending to a horizontal, and then to a descending direction, the rate of decrease gradually diminishing in rapidity as the rate of increase had previously done; so that possibly at the one extreme as at the other a long and almost asymptotical line of convergence or of divergence would be shown.[1]

---

[1] This is in fact what occurs in the case of each coal-pit. At first, after the working of coal has commenced, the returns are small and increase but slowly, the ratio of increase however increases gradually, till at

These being the probabilities of the case on a merely general inspection of the premisses, it is important nevertheless to observe that the period of maximum production, and consequently the beginnings of decay, must be much more imminent if the present law of increase is a *geometrical* one; this theory, in fact, as Prof. Jevons observes, indicates a higher but more evanescent splendour of manufacturing and commercial greatness than the other : it is therefore clearly important to ask: Is our rate of increase geometrical or arithmetical? The exact rate of increase, this question being determined, is at once a more simple and a less important problem.

The figures shown in our table may perhaps be interpreted in either way : it is open to anyone to take the net increase for the whole period, and distribute it in equal proportions over the several years of the period by way of an arithmetical average, amounting on the whole to about 3,250,000 tons per annum. But it is difficult to understand on what principles it can be supposed probable that such a uniform increase, irrespective of the amount of which it is the increase, should take place. Without anticipating later and more detailed discussions, it is obvious that this increase is in a large measure at least relative to and caused by two things, first, the natural increase of our own population, which is a geometrical or proportional increase; and second, the natural growth of capital applied to the development and extension of coal-driven manufactures, which is also unquestionably geometrical. All three indeed are simple examples of the very obvious fact, that all natural non-mechanical increase must show an increased rapidity of growth with the growth of its constituents. Even in those rare cases, therefore, in

last a maximum output is attained ; and the output then gradually slopes down again to smaller and smaller proportions, and at last ceases entirely.

which we find approximately equal amounts of increase, from unequal starting-points, we ought to interpret these as evidence of varying rates of geometrical or proportional increase; the apparent equality is real diversity.

It is clear, also, that whatever law we assume to be the true one for the present, and the immediate future, ought to be capable of being assumed also for the immediate past. Now a geometric decrease backwards from 1854, nowhere seriously contradicts the facts so far as they can be ascertained. Such a calculation would give us about 32,000,000 tons of output in 1832, about 16,000,000 in 1810, and about 13,000,000 at the beginning of the century, which amounts may very fairly be paralleled with the estimates already given. But a calculated arithmetical decrease of 60,000,000 of tons every twenty-two years would bring us to an output of *nil* in 1832, and to *minus* quantities, whatever they may mean in this connection, for all the years before that date.

Prof. Jevons, taking this view of the facts, deduced from the returns of the years from 1854 to 1865, that the actual law of geometrical increase was on the average probably 3·5 per cent. per annum, equivalent to an increase of 41 per cent. in ten years, or 100 per cent. in a little over twenty years. His calculations received striking confirmation in 1871, when the actual production, 117,350,000 tons, differed only fractionally from the calculated amount 117,900,000. The returns for more recent years have not as yet confirmed this estimate: the theoretical amount for 1876 would in fact be about 140,000,000.

But it is obvious that either failure or success in predicting the output of any particular year neither proves in the one case that the theory is erroneous, nor in the other that it is correct. The production is subject to very

violent fluctuations from year to year, varying according
to the tables from an extreme upper limit of increase of
11·2 per cent, to an extreme lower one of — 2·4. The years
since 1873 have been years of almost unexampled depression, and consequently of but languid growth in coal-
production; but a deficit of 6,500,000 below the theoretical amount, is no more than one year's increase in
such years as 1859, 1860, 1861, 1865, 1871, 1872, and
1875 ; so that on the whole calculated period since 1865
the returns are but one year behind, and previous
experience has always shown a period of very rapid
growth immediately succeeding a period of collapse.
The next few years may therefore be watched with
interest as a means of obtaining some more decisive test
of the accuracy of Prof. Jevons' calculations : it is
certain that up to the present, experience has very
largely confirmed them.

If the view of the facts enunciated by Prof. Jevons be
even approximately true, it is at once deducible that our
output will in the next or the immediately succeeding
generation have reached amounts so vast, that we recoil
from them as impossible. But no test in matters of this
nature is so utterly illusory as our incapacity to imagine
the possibility of a thing. It is a perfectly trite observation that the amount of our present output would have
been regarded as something absolutely absurd not so many
years ago; and in fact Prof. Hull himself in 1864, when
the output had already reached to upwards of 90,000,000
of tons, expressed an opinion that the output would never
go beyond 100,000,000.[1]

Our estimates of great or small are mere matters of
mental association and habit; they have no external

---

[1] See Jevons, *The Coal Question*, p. 30, 2nd ed., quoting Mr. Hull in
The Journal of Science, i. p. 30.

validity. If our output is not going to increase in the future according to the law observed up till the present moment, some assignable cause or causes must be given for the variation. This will be the chief matter for discussion in the ensuing chapter.

Before quitting, however, the purely statistical aspect of the subject, there is one further remark to be made. It has been shown that we are practically limited, so far as *exact* knowledge goes, to a period of twenty-three years, or perhaps even a shorter period, if the doubts expressed above respecting the earlier returns are well founded. It is open to any one to argue that this is too narrow a foundation for any very extended superstructure of prediction : that in fact it would be as rational to expect that the ratio of increase as between any two of these years would adequately represent the law of the twenty-two, as that the ratio in the twenty-two can adequately represent the law for the two centuries following. Such an objection would, I think, be valid at least to this extent,—that the limitations of our knowledge up to the present ought to prevent an over-dogmatic insistance on any particular rate of increase, or any consequent conclusion as to an exact period of collapse. It has already been said that the rate of 3·5 per annum adopted by Prof. Jevons has accorded very closely with later experience, but it has not absolutely accorded with it, and if there are any grounds for supposing that the period since 1854 has been exceptional in any important economical respect, this would lead us to reserve our judgment as to the probable rate of increase in the years that are immediately to follow. I make this limitation as to the period with which we have to concern ourselves, because even if the proved rate of increase up to the present goes on for two generations, we shall have reached amounts of output that must inevitably

strain our resources most seriously.   Starting from the
actual output for 1876, the output at intervals of ten years
up to 1936 would be as follows—assuming an increase of
forty per cent. per ten years, which is sufficiently near the
facts for our purpose.

1876	Actual output	133·3	million tons.
1886	Calculated output	186·6	,,
1896	,,	261·2	,,
1906	,,	365·7	,,
1916	,,	512·0	,,
1926	,,	716·8	,,
1936	,,	1003·5	,,

Thus, always assuming that the facts of the last two
decades are adequate indications of the law of the immediate future, we arrive at the conclusion that sixty years
hence our output would be nearly eight times its present
amount, and that something like a quarter of the total
amount of coal existing in these islands at a less depth
than 4,000 feet would by that time be consumed.   Nor is
this all : for it would be utterly misleading to consider this
merely as a quarter of a total every part of which is equally
available.   On the contrary, it would inevitably be the very
choicest of our coal, the most readily available, the most
easily worked, the most intrinsically valuable, which that
quarter would represent; the remaining three-quarters
would be largely composed of seams pronounced unworkable at present prices or anything like them.

Unless, therefore, there is anything exceptional in the
records of the past twenty years, they point to a certain
rise of prices of the most serious character within less than
sixty years; with what effects on our manufactures let the
short experience of 1872 and 1873 teach us.

CHAPTER X.

THE COAL QUESTION.—*Continued.*

IN the last chapter it was shown that the geometrical ratio of increase of about 3·5 per cent. per annum, which the statistics of the last twenty-three years had approximately established to be the law of average growth in the present coal production of this kingdom, could not possibly continue to anything like the theoretical limit at which the whole existing store of coal to a depth of 4,000 feet would be actually raised and consumed. It remains in this chapter to enquire what are the influences which in time must tend to diminish the actual rate of growth, and ultimately convert it into a decrease.

There are perhaps three possible answers to this question. First, it may be said that with the growing perfection of our methods and machinery much of the lamentable waste both in the getting and using of coal will cease, and as more work is obtained from the same or a smaller area of coal in the mine, smaller areas will require to be worked annually, and so the rate of production will decrease. The next view is, that from causes extraneous to our coal production altogether, the immense drain on our coal-resources will diminish; these causes being such as the loss of our old superiority in quality of workmanship, the drunkenness and insubordination of workmen, the closing up of our old markets by protective legislation, and so forth. The third view and, in our opinion, the true one, is that the cause of diminished production will be found in the increased difficulty of working the coal; not absolutely, for there are probably no mechanical difficulties within very wide limits which the ingenuity of our coal workers cannot surmount, but relatively to the greater comparative ease of working other coal-fields, to which ours have been and are still superior, but which are gradually gaining on us as our mines get deeper, and as theirs become more accessible.

The second and third views agree generally in a somewhat despondent view of the future of British industry, they differ in that the former looks to an *immediate* declension, while the latter expects a much greater expansion for a time, conditioned however by an ultimately inevitable collapse when the natural difficulties of coal-raising begin to counteract successfully the mechanical appliances which for a time have overcome them. The former theory, as it gives the chief merit for past greatness to the moral and intellectual superiority of our workmen, so it inclines to attribute the chief blame for future declension to their growing inferiority. The latter considers that on the whole nature has had the chief part in the rise, and will have the chief part also in the fall of Great Britain as a manufacturing nation.

The differences however between these two views are perhaps less obvious than those which separate both from what may be called the optimistic view, which trusts to economies in the getting and using of coal for that reduction in the rate of output of our coal which all who can 'put two and two together' are well aware must be effected somehow, if we are to continue for any great length of time to be a great coal-using nation. We propose therefore to discuss the question chiefly from this side, the bearing of the contrary theories upon it will come out clearly enough in connection therewith.

Let us consider then, first of all, what are our pro-
spects with respect to economies in the *getting* of coal.
It may be remembered that one of the chief heads
of inquiry prescribed to the Coal Commission in 1866
was the question of waste in working, and a large amount
of evidence was taken by a Committee of the Commission
on that subject. In their general report they stated
(vol. i. p. viii.) that, ' although in many instances waste
in working is reduced to a minimum, and although manifest improvement is being made in the working of coal,
especially by the extension of the system of " long wall,"
nevertheless coal is wasted by bad working and by care-
lessness, and that to a very considerable amount in proportion to the amount which is actually used. Under
favourable systems of working the loss is about 10 per
cent., while in a very large number of instances the ordinary waste and loss amounts to 40 per cent.' Whether
however any very considerable proportion of this large
percentage of waste is ever like to be remedied, so as to
reduce the output relatively to the total store and so
postpone the period of exhaustion, is very questionable.
And here a preliminary explanation may be useful. In
almost all discussions upon this subject a radical con-
fusion is apt to appear between what may be called
physical waste, and *commercial* waste. The distinction
may be thus illustrated. If in the coal measures of a particular district there occur within workable depths, say
200 seams of variable thickness and quality, it is extremely probable that not more than half of these would
under present conditions be worked, because no lessee
will work seams at a loss, and if one seam owing to its
thickness and quality is capable of paying 2s. per ton,
while the one above at present prices would involve a
loss, the tendency will be to work out the valuable seam,

even if the less valuable is ruined and rendered for ever
unworkable by the subsidence of the strata. There would
in such a case be undoubtedly *physical* waste, but there
would not be *commercial* waste. As one witness remarked
(vol. ii. p. 329) 'A workable seam is one which will pay
a royalty and provide a profit also for the tenant. If the
public says that a seam of coal which does not fulfil those
conditions should not be left unworked, the public should
consent to pay a higher price for that seam of coal. It is
simply the profit which now decides whether a seam shall
be worked or not.'

I have illustrated this point at some length because
not only in connection with the question of coal getting,
but in other connections also, the truth is fundamental
that what I have called *physical* waste, that is, the wilful
or negligent throwing away of a certain amount of coal,
which might have been made useful had the best known
appliances been adopted, so far from being identical with
commercial waste, that is, the wilful or negligent foregoing of legitimate profits, is often the very opposite.
We may in fact venture to say generally that a nation
which has begun to spare its coal is already ceasing to
be a great coal-using country; so long as it continues
progressive in this direction all improvements tending to
increase the amount of coal as compared with the cost of
getting it, or to decrease the amount of coal required to effect the same amount of work, will instantly be 'discounted,'
so to say, under the influence of competition into cheaper
prices, coupled with extending markets. In the case of
the getting of coal, the tendency to cheapen as methods
of working improve is constantly being counteracted by
the increasing physical difficulties occasioned by our
having to go deeper and further for the coal, so that for a
time there ensues an approximately continuous uniformity

of prices on the average of a number of years, which uniformity we may expect to continue till the limits of mechanical improvement have been attained, that is, till the advantages of superior capital, skill, energy, and situation begin to be counterbalanced by the 'brute' force of more easily worked seams. When at last nature gains the day, as it eventually must, in other words, when economies here cease to cheapen prices, or prevent prices from becoming dearer, and begin to act on the consumption so as to reduce it, the end of the coal-production of Great Britain on the great scale will already have begun.[1]

Practically identical arguments will apply to other economies in the working of coals, such as the introduction of coal-cutting machines, improved methods of sinking and tubbing shafts, cheaper and more effective ventilation, and cheaper rates for labour. It may however be well to enforce what has already been said by noting the facts of the past few years as to two of them: coal-cutting machines and the cost of labour.

It would be outside our purpose to describe in detail the construction of the various coal-cutting machines that have from time to time been invented; all that is necessary is to observe that everyone of them unquestionably economises the coal—that is, makes a larger proportion of the coal of a given area available for use, compared with what is obtained by hand labour. Yet in spite of the panic of 1873, in spite of all the prophecies of coming exhaustion, in spite of the enquiries and protests of the Commissioners, not one of these machines has come into general use, or seems likely soon to

---

[1] So the great corn-exporting countries of the world, are the countries where *wasteful* agriculture is possible. A 'high-farming' country can never, under present conditions, be a great exporter of corn.

do so. And the reason is simply this, that in most cases it pays better at the present price of coal to employ manual labour at the cost of some waste in coal than to sink capital in an expensive machine which would save the coal to the detriment of the coal-owner. Much was heard of these machines when coal was dear; now that it is cheap, we hear no more of them. It is plain, therefore, that coal-cutting machines will come into use only when it pays to use them: and that it cannot pay to use them till a permanent rise in price approximately equal to the rise temporarily made in 1872–73 has been reached. And when they do come into use, the pressure on our manufacturing industries owing to the rise in the price of coal will be so great, that every pound of coal saved by the use of coal-cutting machines will go to lessen the cost of production; in other words, the whole efforts of coal-producers will be directed, not to raising the same coal at less cost, but to raising a greater amount of coal and obtaining an extended market for it by keeping down the price as far as possible. And this will be done with no thought either of the prosperity or the ruin of the country at large, but simply under the influence of competition, and with a view to the direct and individual profit of the coal-owner.

Then again as to the cost of labour. It is plain that the amount of coal *economically* workable depends among other things upon the cost of the labour employed to raise the coal, relatively to the price that can be obtained for it when raised. Diminish the cost of labour, or increase the price relatively obtainable for the coal, and it will pay to work deeper, other things remaining the same. But so far as the price obtainable for the coal is concerned, other things would not remain the same, for a rise in the price of coal would at once affect our manufactures, diminish the advantages enjoyed by this country as compared with other seats of industry, and *pro tanto* hasten the period when it would pay better to manufacture iron and cotton and wool elsewhere.

What then of the other element, the price of labour? Is it possible to calculate on such reductions in this item of the cost of production as may to some extent counteract the growing cost in other respects, and so postpone the date of economic exhaustion? At a first glance, it would seem that a large margin of economy in this respect exists. It is well-known that the scale of wages in our coal-mining industry, as in other industries, is higher than that which prevails in most of the other coal-producing countries of the globe; and in the last few years we have seen how reductions in that rate of wages have been made from time to time, as the price obtainable for the coal became less. But these facts hardly cast any light on the question how far it will be possible to counteract a growing difficulty, not in *selling* the coals but in *getting* them, by reducing the price paid for labour. The experience of 1872–73 seems on the contrary to show that as soon as the demand for coal begins to exceed the powers of our coal-owners to supply it, and a rise in the price of coals ensues, the price of labour tends to rise also. The combined result of these two tendencies would certainly be ultimately to 'drive trade out of the country,' but it is difficult to see what personal motive could be brought to bear upon the miners to induce them meanwhile to take a smaller share of the price obtained for the coal, when *ex hypothesi* that price is increasing, until as in 1873 they are *compelled* to do so, that is, when the mischief is already done. Rising prices for coal mean rising prices also for every article consumed by the miners as well as other people, and a strong individual pressure

is thus being continuously put upon them to exact the
very highest price for their labour that they can get.
Doubtless if labour were as irremovable as the land is, the
stern pressure of necessity would ultimately compel the
miners like everyone else to submit to harder conditions
of living; but so long as emigration is possible, the scale
of wages cannot probably be very largely reduced in this
country below the level obtainable in the United States
or the Colonies without reducing proportionately the
amount of labour obtainable. Moreover ill-paid labour
generally means inefficient labour, as Mr. Mundella has
recently shown in a paper to be referred to immediately;
so that on this ground also it is questionable how far a
reduction in the price of labour would exercise a remedial
effect.

The second great head of influences supposed by some
to tend to reduce production are economies in the *using* of
coal. In discussing the forms and possible effects of such
economies it will be necessary to distinguish between the
various modes of consumption. This has already been
done in another connection; for our present purpose
it will be sufficient if we divide these various modes into
four, whose relative importance may be indicated approximately by the fractions appended to them. These four are :

1. Mining and Metallurgical industries, burning $\frac{15}{40}$ths of our coal annually.
2. Manufactures and locomotion   „ $\frac{11}{40}$ths   „
3. Domestic (including gas and water)   „ $\frac{10}{40}$ths   „
4. Exported coals   taking $\frac{4}{40}$ths   „

In all the first three, enormous economies are still
theoretically possible. So far as the production of power
is concerned this has been already shown, and both in
this and in our metallurgical industries few years within
the century have passed in which some nearer approximation has not been made to the full employment of

the powers contained in coal. The growth of our iron
manufactures has in fact been conditioned by continually
increasing economies in the consumption of coal. The
history of this process of improvement in recent years is
succinctly shown in the following table, extracted from the
Appendix to the recent very valuable paper on ' The Conditions of the Commercial and Manufacturing Supremacy of
Great Britain' read by Mr. Mundella, M.P., before the
Statistical Society, February 19, 1878 :

*Statement of the production of Pig-iron, of Coal used in its Manufacture,
and the Proportion of Coal to each Ton of Pig-iron made in each of
the following years :—*

Year	Pig-iron made	Coal used in manufacture	Coal used to each ton of Pig-iron made
	Tons	Tons	Tons cwts. qrs.
1787	—	—	9 0 0
1840	1,396,400	4,877,000	3 10 0
1869	5,445,757	16,337,371	3 0 0
1872	6,741,929	17,211,729	2 11 0
1873	6,566,451	16,718,562	2 10 3
1874	5,991,408	16,292,201	2 14 2
1875	6,365,462	15,645,774	2 9 0
1876	6,555,997	15,598,381	2 8 0

The immediate purpose for which the above table was
inserted in Mr. Mundella's paper was to controvert, by a
reference to the facts of our recent consumption of coal in
iron manufacture, the law enunciated by Prof. Jevons[1] that
the more economical consumption of coal in the manufacture of iron tends firstly, to cheapen the iron, then and
therefore to extend its use, and so finally to cause more
coal and not less to be consumed in the manufacture of
iron than formerly. It is shown by the table that although
increased economies in the consumption of coal have been
effected yearly between 1872 and 1876, the amount of iron
manufactured has materially decreased, and consequently
the amount of coal consumed in its manufacture.

[1] *The Coal Question*, cap. xv. 2nd ed.

It might perhaps be sufficient in reply to point to the upper portion of the same table, where with a decrease between 1840 and 1869 of half a ton of coals per ton of pig-iron manufactured, the amount of coal consumed in the manufacture of pig-iron was nearly quadrupled.

Nor is it to be thought that Mr. Jevons intended that the effect of coal economies would be such as invariably and immediately to over-ride special circumstances, such as war, famines, panics, national or commercial bankruptcies, and the like, all of which have been exercising their baneful effects since 1873, and on our iron manufacture more than on any other. The iron trade like all other trades is subject to periods of collapse, in which the permanent governing laws of its progress are for a time apparently rendered ineffectual; but on a more extended view of the history of our iron trade in the past, the co-existence of a law of continuous economies in the use of coal, and of continuous increase in the amount of coal used, is absolutely undeniable. Nor is there any difficulty in showing that this co-existence is a case of perfect cause and effect, when once we apprehend the practically unlimited power of expansion of which the trade in an article so universally useful as iron is capable.

Men are always ready enough to doubt whether we are not nearly at the limit of demand when their conversation is of the collective millions of our output; and yet those practically interested in our iron manufactures are quite confident that in their own department they can make a trade, if only they can produce cheaply enough.

To talk as if the demand of the world had reached, or was ever likely to reach a maximum, is to fly in the face of all experience. To take one great industry as an example: in 1850 the railway mileage of the world was about 18,600 miles; in 1865, 90,168 miles; in 1870,

131,641 miles; in 1875, 182,796. It is perfectly certain that the last number no more adequately represents the world's wants than the first did. The total mileage of Europe in 1875 was 88,735 miles, and of America 83,209 miles; while Asia, Africa, and Australia together did not possess more than 11,000 miles. Again even in Europe and America the most enormous disproportion existed between the amounts possessed by different countries, regarded either in respect of area or population, or both, as the following table shows.

Table showing the Extent, Population, and Railway mileage of certain Countries in 1875.

Country	Area in sq. miles	Population	Railway mileage
United Kingdom . . .	121,115	31,000,000	16,626
France	207,900	35,000,000	13,413
Germany	212,091	41,000,000	17,370
Austria	226,406	36,000,000	10,792
Italy	112,852	26,000,000	4,777
Russia in Europe . . .	2,177,864	71,000,000	8,657
Belgium.	11,412	5,000,000	2,162
United States . . .	3,209,844	40,000,000	74,454
Canada	6,534,272	4,000,000	4,106

To argue from the collapse of railway enterprise since 1873 to a supposed cessation of demand is a mere inversion of the facts. As has been several times observed the demand for increased railway accommodation went on continuously up to the moment of collapse; and the collapse was produced net by a cessation of that demand, but by the combined effects of an excessive conversion of capital into fixed as compared with floating capital, and of an extraordinary rise in price, itself the result of a disproportion between the demand and the means of satisfying it. There never yet was any permanent difficulty experienced in selling any amount of iron which our resources can produce, *at a price*; and all economies in its production tend

to bring the cost down to a point at which the world's
price for it can be remunerative.

Every development in this direction tends to multiply
itself, and with every multiplication we may anticipate
further drains on the two great material instruments of
civilisation—iron and coal.

So long therefore as Great Britain continues to share in
the great commerce of the world, so long is it probable that
her manufacture of iron, and with it her consumption of
coal, will increase. It must be admitted that the experience
of the last few years has been otherwise, but there are
abundance of facts in the reckless waste of capital on
foreign loans and other speculative investments abroad
to account for the present great collapse of our export trade.
From the time of the South Sea Bubble downwards there
have been many periods of extravagant speculation, but
none has ever approached the speculative mania from
1870 to 1873 either in the amount, or surpassed it in the
unproductiveness, of its investments. It is calculated that
between fifty and sixty millions of pounds were hopelessly
thrown away during these years in foreign State loans,
and the unproductive foreign investments of other sorts
made during the same period were probably many times
as great. It would be folly to expect that national
extravagance of this character and extent could occur
without the usual results; not only has a large amount
of capital been absolutely lost, but legitimate trade has
been crippled by being diverted from its natural channels,
in purely factitious directions. But there is nothing to
show that the crisis differs from previous ones except in
severity and duration, and anyone of several recognised
possibilities of the immediate future, such as the adoption
in whole or in part of Freetrade principles in America or
the opening up of Africa or Asia Minor, would certainly

or probably have the effect of restoring trade to more than its previous prosperity. It is surely the wildest pessimism to imagine that the progressive activity of the world is likely to suffer more than a temporary check either now or for many centuries to come; and it is quite certain that the conditions which have given Great Britain so large a share of that prosperity in the past, are still present in degree sufficient to justify us in expecting the future to resemble the past *for a time.* But I apprehend that the very recklessness of speculation, the mass and volume of it between 1870 and 1873, are indications of the slow yet unavoidable change from easy supremacy to supremacy under conditions of eager and growing competition. The influences which induced such enormous investments of floating capital abroad must in time induce the transfer of other capital as well, but the process will be a slow and difficult one, and I anticipate that the *immediate* effects will be rather to increase the amount of our industrial activities, and so enhance our present rate of coal consumption, even though this must eventually involve us in a more rapid exhaustion. One example of this tendency already noticeable will be discussed later on in connection with the subject of our coal exports.

If the above necessarily vague and imperfect considerations are justified, it is quite certain that in the long run all economies, whether of coal or anything else, in the production of iron, will tend to increase the amount of iron which it will pay the world to use, and so to increase the amount of coal used in the manufacture of iron. So long as from whatever causes it pays best to purchase that iron from us, so long will the drain on our coal-mines continue; when our coal becomes scarce and dear, the world will not to suit our convenience reduce the amount of iron it requires, but will go and buy it elsewhere. Of course the process will be gradual; those branches of iron manufacture into which the cost of the raw material enters most largely will probably go first, and the struggle will doubtless be a prolonged one, but the whole question is practically settled when it is admitted that we have competitors, and that the drain on our resources tends to make their position gradually more favourable.

The second great head of consumption, that for our manufactures and locomotion, need not detain us long. Practically the same laws must hold for this as for our iron manufactures. We can only continue to be the great purveyors of manufactured goods in the world's markets by being prepared to push our goods wherever and whenever an opportunity offers. It is absolutely necessary therefore that every increased advantage in the way of greater economies of motive power shall be converted into capital to be devoted to further extensions; and if one manufacturer refuses to do so, he simply transfers those additional profits through the agency of his bankers, to others who are eager to do it in his stead.

With respect to locomotion there is one special fact to be noted. From a variety of causes, but more especially from our freedom from war, our unequalled maritime position, our colonial connections, and our still unapproached capacities of coal and iron production, we have been becoming with every year more and more the carriers of the world. Thus the tonnage of British vessels entered and cleared at ports in the United Kingdom increased between 1860 and 1870 from 13,900,000 tons to 25,000,000 while the foreign tonnage had increased from 10,700,000, to 11,500,000 only; the amounts for 1876 are, of British vessels 33,500,000; foreign, 17,300,000 or little more than half. This increase is chiefly in our steam navy, which has increased from 500,000 tons in 1860

to nearly five times that amount in 1876, or to a total
tonnage twice as great as that of all the other steam
merchant navies of the world combined.[1]

So long as we retain the advantages above enumerated,
so long we may expect our supremacy in this respect to
continue, and in an increasing ratio; but the very increase
in our consumption of coals both in the construction and
propulsion of such a navy must bring the time more
rapidly near when one at least of those advantages and
an absolutely essential one, cheap coal, will no longer be
ours. Here again we can see no limit but a rise in price,
the product of growing scarcity of fuel.

We pass next to our consumption of coal for domestic
purposes. To this alone, as we have said, about one fourth
of our total production, or 33,000,000 tons per annum
are applied. It has been pointed out by one writer after
another that our consumption of coals for these purposes
is an eminently wasteful one, and that we obtain in our
open fires a mere fraction of the heat which this amount
of coal scientifically consumed might be made to produce.
Yet the effects of such teachings have been extremely
small even as to individuals, while on the mass of the
people they have had no effect whatever. Our domestic
consumption grows with our population in a ratio which
rather increases than diminishes, although we may describe it on an average at about one ton per annum per
head of population. The excessive prices of 1873 revived
the public interest for a time in more economical methods
of consumption; and the effect in the way of reduction of
consumption was immediate. In London, where considerable improvements were probably effected at that time
a small diminution in the rate per head of population

[1] See Mr. Farrar's article on The Strength of England, *Fortnightly
Review*, March, 1878.

seems to have resulted; but the decrease is after all inconsiderable, and we cannot certainly say how far the prevailing commercial depression may not entirely account for it.

But this example only shows that in this case as in the others nothing is likely to interfere with the continuous growth also of this item of consumption, except a permanent and serious rise in price. Every year of commercial prosperity extends the numbers of what may be called the 'comfortable classes,' and extends also their ideas of what comfort is ; so that every year the consumption of luxuries, and among the rest the luxury of a good and cheerful fire, tends to increase.

Here also therefore we are face to face with the same difficulty; the condition of our prosperity is abundant and cheap coal, the conditions on which alone we can continue long to possess abundant and cheap coal, so far as our voluntary action is concerned, will not or cannot be realised, save under the pressure of dearth and scarcity. The indefinite continuance of our commercial prosperity in fact involves a contradiction in terms.

Special emphasis has occasionally been put upon the possibility of a very material reduction of our consumption by the invention of other modes of lighting, more particularly by the introduction of the electric light. It is not to be denied that could a light be obtained as efficient, convenient, and cheap as coal-gas, its introduction would in the first place tend to reduce our consumption of coal very considerably, for about sixty tons out of every thousand, or in the whole about 8,000,000 of tons annually, are converted into coal-gas in this country.

This admission however must be made with several provisoes : in the first place, the new materials must not involve an expenditure of coal in their preparation comparable with that which is saved by the abolition of our coal-gas manufacture. Whether this proviso can be satisfied, others must decide. Electricity for practical use is not obtained from the clouds or the air; it has to be manufactured, and in its manufacture we must draw upon some store of energy already existing, since energy is no more capable of being created than matter is. It is highly improbable that any great source of energy, as distinguished from a mere variety in its form, is still unknown, and the course of events has sufficiently settled that wherever it is possible the store contained in our coal is sure to be drawn upon, in whatever form it is intended to be used. Coal would be used in the reduction of the metals, in the preparation of the acids, in the construction and driving of the machines; and when we consider that at present it is far more expensive to obtain useful work by converting the energy of heat into that of an electrical current than to obtain it at once from the heat itself, it seems highly improbable that for ordinary commercial purposes electricity will ever supplant coal-gas, without still involving a very considerable expenditure of coals.

Even however should it be found more economical so far as coal is concerned, to obtain light from electricity than from burning gas, two questions still remain: First, whether the increased advantages of such a mode of lighting would not tend to such an increase in its use as would fully counterbalance any diminution per unit of light in the amount of coal required to obtain it? There is no reason why our streets and roads should not be made as bright as day at midnight except the cost; and when we compare our own ideas of lighting with those of a hundred years ago, the suggestion of the possibility of some such advance in the future does not seem so very extravagant.

Secondly, if any portion of the assumed saving is not

expended in this way, it is very certain that it will be expended in some other; by whatever sum a manufacturer's lighting charges were reduced, by the same sum would his productive resources be increased, and this again ultimately means more coal.

From every point then it would seem that we reach this fact that our coal trade is one which developes itself according to laws that individually we are perfectly powerless to affect; if it seems to promise a less rapid increase here, it is only that it may spread abroad with accelerated vigour elsewhere; if it is our slave in some aspects, it seems as if it were master in others.

Finally, we have to ask What of our export of coals? Is the same law of growth visible there, and what is the prospect for the future? The record of that growth was shown in diagram 1 (p. 310), and in the accompanying table; and to these we must now again refer. It is there seen that rapid as has been the growth of our total production during the last twenty-three years, the growth of our export of coals has been greater still. Beginning with 4,300,000 tons in 1854, we find it reaching 16,250,000 in 1876, that is, it has nearly quadrupled. As a consequence its proportion to the total production has risen from about one-fifteenth or 6·7 per cent. to close upon one-eighth, or 12·2 per cent. At such a rate of increase it would seem as if our whole annual production would ultimately be swallowed up in our exports; and it is perhaps not impossible that even after we have ceased to be to any great extent a manufacturing people, a certain export trade in coal may still continue : just as the export trade in coal preceded by centuries our own uses for it other than domestic, so it may also survive these by a period as prolonged.

To account for this persistent and growing export

trade, we have to consider, first, the special advantages, which have contributed to make England hitherto the great coal-producing and coal-exporting country; and second, the circumstances which have led foreign countries increasingly to take our coals: first, the circumstances which have made us *sellers*, and second, the circumstances which have made them *buyers*.

To the first many circumstances have contributed. There are the generally excellent quality of our coals, the regularity of their disposition, and the comparative ease with which they can be worked;[1] there are the energy and perseverance of our people, and their hereditary capacity for marine adventure; there are the order, good government, and freedom, which have combined with a comparatively enlightened policy in favour of commerce and manufactures to make England in all ages a home for industry and enterprise. But all these together could hardly have made England for five centuries the purveyor of coals to the world had it not been for the extraordinarily favourable position of her chief coal-fields for marine conveyance. A glance at a map of the coal-fields of the world would show that nowhere else can coals be so easily and cheaply put on board ship as they can in our Newcastle, our South Wales, our Cumberland, and Scottish Coal-fields.

The only places comparable in this respect are isolated

---

[1] Cf. *Coal and Mining* (p. 48), by W. W. Smyth, Esq., F.R.S., where in speaking of the Newcastle Coal-field he enumerates the following advantages: first, 'the general regularity of the measures, dipping at a very moderate angle, commonly about 1 in 20; the convenient thickness of the seams, from 3 to 6 feet; the excellent qualities of the coals, and the usual goodness of the roof, which allows of wide working places and roads, with a very small expenditure of timber.' It is only in speaking of the *newer* pits that he has any disadvantages to mention. And for the advantages of the South Wales coal-field, see *ibid.* p. 68. And contrast the French coal-fields, *ibid.* p. 75; the Belgian, *ibid.* pp. 77, 78, the Russian, *ibid.* p. 86.

coal-fields such as those of Australia or Labuan; within the limits of our older civilisation the combined advantages for coal-transit to foreign ports enjoyed by Great Britain are quite unparalleled. Till within a quarter of a century of the present time, these advantages gave Britain the practical monopoly for her coal of all the coasts of Europe; and America even was then a considerable importer. It was the invention of railways, themselves the products to a large extent of our coal, which gave the first great impulse to the coal-production of other countries.

The vital importance of means of cheap transit for the development of the coal-fields of any country, was illustrated a century before in England in the effects of the great canal system which then began to be introduced. Thus the development of the Lancashire coal-field dates from the formation of the Bridgwater Canal in 1765; that of the Yorkshire coal-field from the formation of the Leeds and Liverpool Canal about 1780; that of the Derbyshire coal-field from the formation of the Erewash Canal about the same date. Between 1750 and 1780 the royal assent was given to Acts authorising no less than thirty-six distinct canals in Yorkshire, Nottingham, and Derby; and the total number of canals for England during the same period amounts to about a hundred. Acts for forty-four additional canals were obtained between 1800 and 1813.[1]

The chief purpose of the great majority of these canals was to obtain access to a market for inland coals; and the effects were often very great. Thus on the Leeds and Liverpool Canal there were carried in 1783 about 86,000 tons of coal to Liverpool; in 1793, 252,500 tons.[2]

But great as were the effects of the introduction of

[1] See Coal Commission Report 1871 : iii. app. nos. 22, 25.
[2] See Coal Commission Report, iii. 15.

canals in this country, the effects of the introduction of
railways have been vastly greater. This is well illustrated
by the following table of the quantities of coal imported
into London for certain years before and after the
development of the metropolitan railway system.[1]

Table showing the quantities of Coal imported into London.

Year	Coastwise	Canal	Total
1837	2,626,997 tons	2,324 tons	2,629,321
		Ry. and Canal	
1845	3,392,512 „	68,687 tons	3,461,199
1855	3,016,869 „	1,161,086 „	4,177,955
1865	3,161,613 „	2,741,588 „	5,903,271
1875	3,134,846 „	5,076,046 „	8,204,892
1877	3,170,601 „	5,421,081 „	8,591,682

It is to be noted that the sea-borne coals to London
have not increased, have even slightly diminished in the
last thirty years, while an entirely new inland trade has
developed from practically *nil* to the gigantic amount of
5,500,000 tons.

The effects of railway extension are also seen in the
following table :

*Table showing approximately the output of important British Coal-fields
in 1855 and 1875, with percentage of increase.*

Coal-fields	1855	1875	Percentage of increase
Northumberland and Durham	15·5 mill. tons	32·3 mill. tons	108
Cumberland	0·8 „	1·17 „	50
Yorkshire	7·75 „	15·86 „	118
Nottinghamshire . . .	0·8 „	3·25 „	306
Derbyshire	2·25 „	7·19 „	220
Warwickshire . . .	0·25 „	0·8 „	220
Leicestershire . . .	0·42 „	1·17 „	179
Staffordsh. and Worcestersh. .	7·32 „	14·5 „	98
Lancashire	8·95 „	17·9 „	100
Cheshire	0·75 „	0·69 „	− 8 [2]
Shropshire	1·43 „	1·92 „	34
N. Wales	1·12 „	2·35 „	110
S. Wales	8·55 „	14·17 „	65
Scotland	7·32 „	18·69 „	154
Total for United Kingdom .	64·45 „	131·8 „	105

[1] Condensed from the *Mineral Statistics.*   [2] Decrease.

341 THE COAL QUESTION.

It will be observed that the coal-fields which show abnormal increase are entirely inland; a large proportion of the Scottish increase here thrown together is to be attributed to the development of the inland coal-fields of Ayrshire, Lanark, and the Borders.

If we now compare this table with the table next following which shows the approximate outputs for the same years of the chief coal-producing countries of the world, a similar law of increase will be noticed.

Table showing the approximate outputs of the chief coal-producing countries of the world in 1855 and 1875.

Countries	1855	1875	Inc. per cent.
United Kingdom . . .	66·0 mill.tons	132· mill. tons	100
United States of America .	5·0 ,,	50· ,,	900
Germany	8·0 ,,	50· ,,	462
France	7·5 ,,	17· ,,	126
Belgium	8· ,,	15· ,,	88
Austria	2·5 ,,	12· ,,	380
Russia	·5 ,,	1·5 ,,	200
Australia	·25 ,,	1·5 ,,	500
Nova Scotia, &c. . . .	·5 ,,	1·5 ,,	200
Spain	·25 ,,	·75 ,,	200
India	·5 ,,	·5 ,,	0
Other Countries . . .	2· (?) ,,	2· (?) ,,	0 (?)
Total .	101· ,,	278·75 ,,	176

* For later returns, See Appendix.

In several cases the rate of increase has been vastly accelerated by particular circumstances. Thus in America there has been an abnormal increase owing to immigration; whether this influence will continue with equal strength in the future it is impossible to say. Then in America, in Germany, and in Austria, the period has been very largely a period of transition from wood to coal. The great forests in these countries have been rapidly reduced during the last twenty years; and in America the hiccory, pine, and oak fires of the earlier period have been almost

entirely superseded by the anthracite stoves. It is highly
probable therefore that the rate of growth in future years
will be less rapid on this account than it has been.

Still, enough remains for which railway extensions
can alone account. In all these countries the coal-fields
lie at very considerable distances from the sea, and access
to other parts of the country was in many cases rendered
difficult by intervening mountains, or other physical obstacles; but with the introduction of railways these
obstacles have been practically removed, and, as the table
will show, the coal production, more especially of America
and Germany, has developed and is developing with
enormous rapidity. Nor need we anticipate any check
upon this development in either country, certainly not in
America, from limitation of the supplies. It is impossible
as yet to obtain statistics on the subject which will make
a definite comparison with our own coal-fields possible;
but it cannot be doubted that in both countries enormous,
and in America practically inexhaustible supplies exist.[1]

It is difficult at first sight to reconcile these facts with
the aspects of our own coal export shown in the table on
the opposite page.

It is there shown that the two chief importers of our
coals are themselves great coal-producers, one of them
the greatest coal-producing country on the continent, and
the one which has grown most rapidly. But a somewhat
closer investigation solves the difficulty. In the Appendix
to the report of Committee E., Coal Commission Report
Vol. III. pp. 113 seq., the ports to which our coals were
sent in the several countries are shown in detail. These
tables show us that our coals are distributed over every

---

[1] For further information as to these, see Coal Commission Report iii.,
pp. 206 *seq.*; W. W. Smyth's *Coals and Coal-mining*, caps. vii. viii. ix.;
and Rogers' *Coal-fields of Pennsylvania.*

Table showing the destinations and amounts in 1856 and 1875 of British Coal exported.[1]

	1856	1875	Inc. per cent.
France	1,136,299	2,610,865	130
Germany	860,877	2,172,913	152
Sweden and Norway . .	236,116	1,136,059	381
Italy	143,753	987,868	593
Russia	213,553	895,195	319
Denmark	402,875	749,399	86
Spain	182,339	648,650	254
Holland	196,220	454,019	131
Belgium	37,739 (?)*	324,682	760 (?) *
Austria	65,780	73,547	11
India	208,319	602,914	189
N. America	309,499	234,422	−25
S. America	206,154	874,864	324
W. Indies	164,225	467,775	185
Total Export	5,879,779	14,475,036	146

* The Belgian import for 1856 is exceptionally low.

section of the coast of Europe. Thus in 1869, 133 ports in France alone received cargoes of our coals in amounts for the year varying from 50 tons up to 100,000. The great ports, Havre, Bordeaux, Dieppe, Dunkirk, Boulogne, Calais, Rouen, received amounts from 100,000 to 64,000; but every little fishing village seems to have purchased some. So in Germany and Prussia we find 64 ports; in Denmark 120; in Sweden and Norway 97; and so with every other country.

Practically it comes to this, that in spite of the great extensions of the continental railway system effected during the last twenty years, the greater portion of the continental coasts is still commercially nearer our coal-fields than their own. All round the coasts of Europe therefore we are feeding the smithy fires of their villages, lighting their streets, and warming their houses, supply-

[1] Compiled from the *Mineral Statistics*, and Coal Commission Report, iii. 1871.

ing their arsenals and their ships with fuel, and contri-
buting steadily in these and similar ways to the growth of
their prosperity.

This process has been looked upon with increasing
suspicion by many thinking persons in this country. It
is perfectly obvious that such a trade as this, which is an
exhaustive process not a recuperative one, must have
more cogent grounds to justify itself than the ordinary
common-places of Free-trade. And here again we are
met by the dilemma : How far we are to sacrifice the
present with a view to the future. If no more hangs on
our export of coals than the present loss of the price we
obtain for it, together with the profit of freight for its
exporters, and if we are quite sure that we may realise
both of these at any future time which we may choose,
then doubtless it might be well to place such a duty upon
its export as will have to some extent a prohibitive effect.

But it is impossible to suppose that the results of such
a measure would be anything like so simple.   Prof.
Jevons in the very interesting chapter on this subject in
his work on the Coal Question, urged that such a duty
would be practically an impost on our manufactures, inasmuch as the export trade in coal was the correlative of
our. import trade in raw materials.   That in fact the
facility with which importers of bulky raw materials
could obtain a return cargo which would certainly pay
its own expenses, acted directly in reduction of home
freights, and so gave our manufacturers an advantage
over all other importers for manufacturing purposes.[1]

There can be no question that to a certain extent this
is true ; but the evidence seems certainly to show that
the importance of this aspect of the question has been
exaggerated.   France, Germany, Russia, Sweden and Nor-

[1] See Jevons, *The Coal Question*, chap. xiii.

way, Italy, and Denmark take between them every year *five-eighths* of our whole enormous export of coal; or a total in 1876 of close upon 10,000,000 tons. Now our great manufactures, other than metals and machinery, for which we are not materially indebted to anyone, are of course our cotton, wool, and linen. The raw materials of these together with dye-stuffs cost us yearly about 100,000,000*l.*, and the value of the goods exported is somewhat over the same amount, in 1874 about 115,000,000*l.* Now these six countries send us none of the cotton, just one-fifth of the wool, less than one half of the flax, and one-tenth of the dyes; in all out of the 100,000,000*l.* worth, just one-tenth in value. Our great sources for all of them are the United States which take scarcely any of our coal, and India and Australia which take very little.

That the relation between our imports of raw materials and our export of coal is in respect of a large portion of it unimportant, is seen also from this, that in the case of Newcastle which exports upwards of 4,000,000 tons, the steam colliers (as I have been informed) invariably return in water ballast, the others usually loaded with chalk or other similar materials. Our coal-trade in fact tends very largely in some quarters to become a coal-trade pure and simple, with no direct relations whatever to our other commerce.

Of course all articles imported from those countries which import our coals are *indirectly* useful to us as a manufacturing people, corn for example and timber. But these would hardly seem to be within the meaning of Mr. Jevons' argument. It by no means follows however that even in the case of the export trade pure and simple, it would be judicious to impose a prohibitive duty. Apart from the considerations urged with great force by

Prof. Jevons, which would render such a measure of doubtful expediency on general grounds, such as our commercial engagements with other countries, our position as the champion of Free-trade and the like, it seems exceedingly doubtful whether matters have not reached such a point in this country, that an export-duty would tend rather to increase the cost of coal at home, and consequently diminish the amount of coal *economically* workable; so that here again we should, in our anxiety to *save* our coal, really *lose* it for all practical purposes.

The way in which I conceive this impost would have the effect above indicated is as follows:—Under the conditions of coal-getting now prevailing in England, it is impossible for it to continue to be a profitable business, unless it be conducted on a very extensive scale. There has been a constant process going on of elimination of the smaller collieries, and substitution for them of larger concerns, working a more extensive area under one management. Now it is perfectly impossible that such immense undertakings, with their extensive and costly plant, should pay, unless a tolerably uniform sale for the coals raised can be secured. For this purpose it is impossible to trust to the home consumption alone; so far as the metallurgical and manufacturing industries are concerned, the most violent fluctuations inevitably occur; the domestic consumption is much less variable, but it too is sensitive to trade depressions; the ' compensation balance ' or ' governor' so to speak in respect of these fluctuations has been and is the export trade. It is spread over areas so widely separate, that the influences of dull trade reach them more slowly and incompletely; it is therefore possible to ' force a trade' abroad, when times are dull at home. Thus if reference be made once more to the table of production and export (page 310), it will be found that not infrequently the *earlier* years at least of a decreasing production have been years of great activity in export; compare more particularly the years 1855–7, 1861–2, 1866, 1870, 1874, 1876.

If this deduction be a just one, it is perfectly obvious that a stop artificially imposed upon this means of relief would render coal-getting an increasingly precarious and unprofitable business; seams that might otherwise have been worked at a profit would never be opened or would be abandoned, and the period when it would not pay to work the coals at all would be brought proportionately nearer.

Such a check must also, of course, act immediately upon our manufactures; coals continue cheap because of the continued influx of additional capital into the business; without this influx coals would increase in price; the hard race we even now run with foreign competitors would become harder, and in this way also the time would be hastened when we should cease to require our present enormous output, and when therefore it would cease to be raised. The conclusion of the whole matter would in fact appear to be, that on one condition alone can we continue to enjoy in the present the manifold advantages of cheap coal, viz., that we *raise and get rid of our coal as fast and as freely as possible.* And when we consider that the motives for keeping coal cheap now are personal and pecuniary, affecting every coal-owner's, every manufacturer's pocket here and now, while the motives for sacrificing the present to a possible future, are remote, contingent, and sentimental, there can be little doubt which will gain the day in the future as in the past. Legal or other pressure might do harm to existing interests; it is very doubtful whether it could postpone one hour the inevitable day when the long race so bravely run must go against us.

When that day comes, it matters not what numbers of millions of tons may remain within 'workable' depths, our coal-fields will be as truly exhausted for the purposes of a great industrial country such as Great Britain is at present, as if the whole had absolutely vanished. As was hinted in the previous chapter, it does not at all follow that we shall not continue to hold an important place in the family of nations; our land is not without natural fertility, and the humidity of the climate gives it a very considerable advantage in textile manufactures. It is very possible, therefore, that with a scantier population and more frugal ways, the coal that remains to us, or the stores of energy which we possess in our abundant streams may (as in Switzerland) retain for us a not unimportant place as manufacturers of special fabrics. Our unrivalled maritime position must long make our country also a suitable centre for a great carrying trade; in this way it is difficult to see why we should not emulate the prosperity of Holland, which possesses little or no coal.

If only our descent from our present extraordinary position be a gradual one, much may be done in the interval to adapt ourselves to more unfavourable conditions; and as has been already said, the limits of possible economies in the way of personal indulgence are so vast in this favoured country, that many great diminutions in the cost of labour and in rents, may be made with a view to stave off defeat. But it is very certain that nothing of this sort will be submitted to save under the stern persuasion of necessity, in other words until the sceptre which cheap coal bestowed is already passing away from us.

When these things will be, soon or late, is a very different matter; he would be a rash speculator who fixed

himself to a year, or a generation. The enormous accumulated capital of this country, its resolute respect for the
public credit, its order and good government, and the
unequalled efficiency of its labour, are advantages which
can hardly be exaggerated. They will certainly help to
prolong the struggle, but they can hardly postpone defeat
indefinitely. They may make our decline more gradual
and less bitter, but they cannot for ever exempt us from
the inevitable.

Being inevitable, the best philosophy for us is to find
what of good is contained in it. Sentimental regrets that
these very hills and valleys will no longer resound with
the din of labour, or be blackened by the smoke of forge
and factory, would surely be out of place. What we might
regret would be if the Britain which we know and are
proud of, the Britain of great achievements in politics and
literature, of free thought and self-respecting obedience,
of a thousand years of high endeavour and constant progress, were indeed to perish, when these factories and
furnaces whirled and blazed their last. But it is not so.
This country's fortunes are gradually being merged in
those of a greater Britain, which, largely through the aid
of the coal whose prospective loss we are lamenting, has
grown beyond the limits of these islands to overspread the
vastest and richest regions of the earth; nor have we any
reason to fear that the great inheritance which America
and Australia and New Zealand have thus accepted from
us will, in their hands, be dealt with unworthily.

APPENDIX.

THE returns of British coal production for 1877 are not yet complete. The coal *export* was 15,358,828 tons; and we gather from the inspectors' reports that an increase in production of about 54,000 tons only has occurred.

The following are a few later returns of foreign coal production:

United States	. . .	1877	54,398,250 tons
Germany	. . . .	1877	48,337,950 „
Belgium	. . . .	1876	14,329,578 „

INDEX.

MED

Medullary rays in calamites, 78
Medullary rays in the lepidodendron stem, 93, 98
Megalichthys, fossil ganoid fish, 146, 147
Meldrum ; his share in the 'Paraffin Industry,' 220
Menobranchus, a recent amphibian, 114, 126
Meteorological changes, their effect on fire-damp, 190
Microscope, its importance in geological research, 13, 17, 19, 21, 24, 107
Microscopores. *See* Spores
Middle coal measures. *See* Coal Measures
Miller, Hugh ; scenery of the Coal Measures, 127
— fossil dipterus, 138
Millstone Grit, 35, 50, 53, 56, 59, 68
Mineral charcoal, or 'Mother of coal.' *See* Charcoal, mineral.
Mineral springs, 65, 66
'Mineral statistics of the United Kingdom,' 309
Mitscherlich ; discovery of nitrobenzene, 212
Mollusca, fossil remains of in coal measures, 50, 119, 151, 162
Monocotyledons in coal, 108
Morasses. *See* Swamps.
Mosses, their absence in coal measures, 108
— chemical composition of, 165, 169. *See* Club Moss.
'Mother of Coal.' *See* Charcoal, mineral.
Mud, nature and geological action of, 7, 8, 9, 12, 13, 18, 26, 31, 49, 56
Mundella, W., M.P. ; on coal statistics, 328
Münster, on the animals of the coal measures, 111
Murchison, Sir Roderick, F.R.S.; statistics of coal supply, 305
Murdoch, W.; production of coal gas, 201, 202
Myriapoda in coal measures, 154, 157

NAPHTHA, Napthalin, 212
Nasal passages of recent and fossil fishes, 141
New Red Sandstone. *See* Sandstone.

PER

Newcastle, its trade in coal in 1281, 227
— its trade with London prohibited by Charles I., 229
'Newe-castele coles' exported to France, temp. Edward VI., 229
Nicaragua, lake of, sharks and dogfish in the, 134
Nitrobenzene, 212
Northumberland, geology of, 39
Nova Scotia, coals of, 17, 19
— Myriapods in coal measures, 157

OAK, chemical composition of the wood of, 165
— axes or picks of, found in coal, 226
Oil, paraffin. *See* Paraffin.
Old Red Sandstone. *See* Sandstone.
Oolite of Russia, animal fossils in the, 133
Oolitic coal, 25
Ophiderpeton, 126
Ophioglossaceæ in coal, 108
Ophiomospha, 126
Otto and Langen's gas-engine, 280
Owen, Prof.; on mastodonsaurus, 112, 116
— on lepidosiren, 141

PALÆONISCUS ; ganoid fishes in coal, 148
Paraffin oil and other products; origin of the 'Paraffin Industry,' 220
— processes of production and purification, 221
Parasphenoid, 133
Paris, English coal exported to ; its use prohibited, 228
Peat, chemical composition of, 172, 176
— production of paraffin from, 222
Peat-bogs, 57
— — in Ireland, 172
Pebbles, 8, 15, 47, 54
Pecopteris, in coal, 108
Pennsylvania, deposits of anthracite, 175
— gaseous exhalations from oil wells, 178
— petroleum springs, 222
Perkin ; aniline and aniline colours, 213

Spottiswoode & Co., Printers, New-street Square, London.

www.ingramcontent.com/pod-product-compliance
Lightning Source LLC
Chambersburg PA
CBHW021358210326

41599CB00011B/930